The Martian Landscape

National Aeronautics and Space Administration

NASA SP-425

By the Viking Lander Imaging Team

Scientific and Technical Information Office
NATIONAL AERONAUTICS AND SPACE ADMINISTRATION
Washington, D.C. 1978

Foreword

Not long ago the idea of taking pictures of Mars from its surface was an idea located intermediately between far out and preposterous. It changed from a dream to a concept about ten years ago with the advent of the Viking Program. As so often happens in the exploration of space, we were able to push back the boundaries of the practicable. In retrospect, with success under our belts, it even sounds simple: put cameras on spacecraft, land them on Mars, take pictures, send them back to Earth. Were it so!

In this book Tim Mutch, leader of the Viking Lander Imaging Team, takes you on a journey spanning a decade. Suffer with him as he copes with innumerable meetings, arguments, alternate designs, budget problems, incipient failures, and, at times, sheer exhaustion. Enjoy with him amazement at how teamwork and dedication can manage the impossible. Share in wonderment at technical intricacy, the occasional euphoria of success along the way, and the final exhilaration when magnificent photographs flow back from the rocky plains of Mars.

The Martian Landscape is a tribute to the hundreds of skillful people who made Viking happen. Thanks to them, you are there.

April 1978

Noel W. Hinners
Associate Administrator for Space Science
National Aeronautics and Space Administration

Contents

The Viking Lander Imaging Investigation: An Anecdotal Account

Thomas A. Mutch

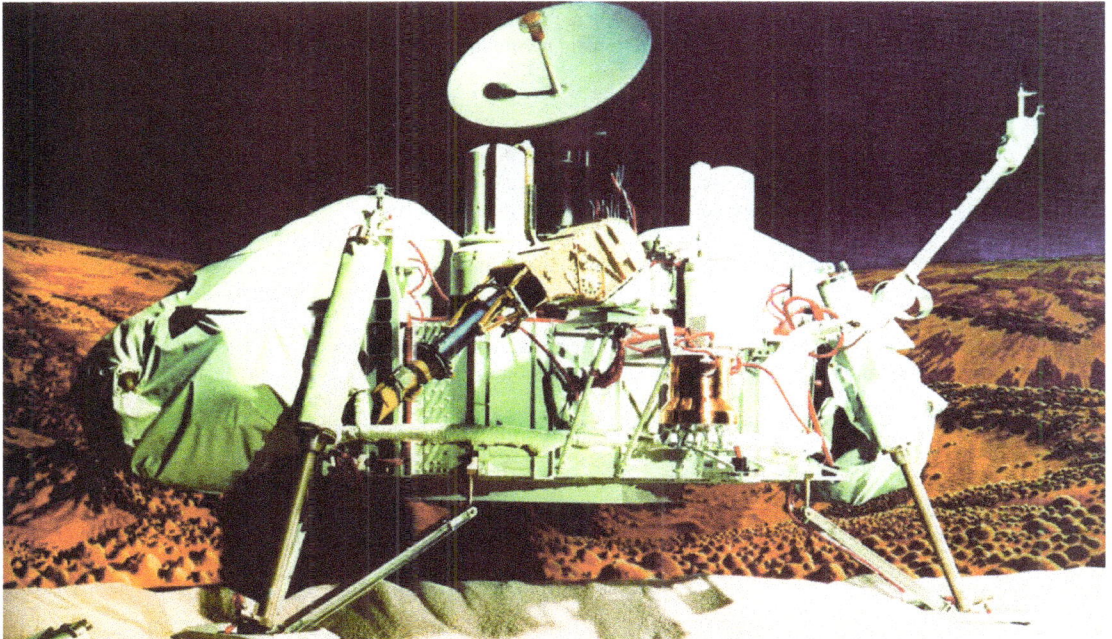

The Final Test

It is the morning of July 20, 1976. After years of endless work and unrelenting deadlines the last night has been a strangely peaceful interlude.

For a month now the Viking spacecraft has been circling Mars, 360 million kilometers from Earth. Yesterday the Lander was coupled to the Orbiter. The onboard computers were loaded with instructions for separation and landing. Now they are carrying out those instructions, insensitive to further advice from Earth.

At 1:51 a.m. the Lander separates from the Orbiter and begins its descent to the Martian surface. Approximately at the same time I drive through the cool California night to the Jet Propulsion Laboratory. The windows of the tall buildings sparkle with lights. The parking lots are full. People hurry past in the darkness. I walk quickly to the building where the Lander Imaging Team is housed. Many of my colleagues, scientists and engineers, are there. For all of us there is only waiting, and I realize that I would rather wait alone, away from forced conversation. I walk to a nearby building and take my assigned position in the "Blue Room," a broadcasting area where the first pictures will be received and transmitted to the news media assembled in an auditorium.

5 a.m. The final descent begins. Conversation stops – an overwhelming silence. We listen to the mission controllers as they call out each event. After years of waiting, hoping, guessing, the end rushes toward us – too fast to reflect, too fast to understand.

5:05 a.m.   "400 000 feet"
5:09 a.m.   "74 000 feet"
5:11:43 a.m. "2600 feet"
5:12:07 a.m. "Touchdown. We have touchdown."

It worked! Amazingly, it worked. Everywhere people are cheering, shaking hands, embracing. I decide not to join the celebration. It is too soon. Forty minutes more remain before the first picture from the surface of a far planet will assemble on the television screen.

5:54 a.m. I study the blankness of the television screen, waiting for the narrow strip of light that will signal the first few lines of the first picture. And it appears. A sliver of electronic magic. Areas of brightness and darkness. The picture begins to fill the screen. Rocks and sand are visible and – finally, at the far right – one of the spacecraft foot pads, a symbolic artifact that stamps our accomplishment with the sign of reality.

I wait impatiently for the second picture, a 300° panorama looking out toward the horizon. On Mars the camera carried out its slow, arcing traverse minutes ago. Now rock-strewn ridges, drifts of sand, distant bluffs slowly pass before me.

All this time I critique, for the audience watching elsewhere, the landscape we are viewing. It is not a task I have been looking forward to. But now excitement washes over my inhibitions.

Time and time again I repeat, "It's incredible." And it truly is. Nothing before or after can compare. It is transparent, brilliant, boundless. An explorer would understand. We have stood on the surface of Mars.

6:52 a.m. The first two pictures end. The Orbiter, which has been relaying these first images to Earth, drops below the horizon, and the Lander prepares for its first night on the surface of Mars. On Earth, we plan for the days ahead.

The Beginnings

We live in an age with little patience for history. I was frequently reminded of that in the first few days after our successful landing on Mars. Continually I was asked, "What are your thoughts as you look at these pictures?" What were my thoughts? A kaleidoscope of memories eight years of planning, moments of frustration, friendships forged by common problems, and now everything happening just as we had disbelievingly promised each other it would.

The first few times I was asked about my thoughts I tried to describe those eight years embedded in the first picture. And that was when I discovered that history was not the subject of the hour. Quickly enough I learned to give the desired response, a crisp geologic description sprinkled with superlatives, sized to fit a 30 sec spot on tomorrow's news program. But I continue to think about the history. If you want to appreciate these pictures fully, you have to travel with us as the Viking Project is transformed slowly and painfully from an idea to a durable spacecraft, propelled on its long journey to Mars.

The Viking Mission was first defined by NASA in 1968. Its predecessor, Voyager, never passed beyond the talking stage. Starting in 1965 and continuing through 1967, tentative plans had been developed for an integrated long term program of Martian exploration involving, first, flyby and orbiter missions, and then a series of lander missions in 1973, 1975, and 1977. Each of these Voyager Landers was to be launched by a giant Saturn V rocket. Successive missions were to contain increasingly sophisticated scientific equipment, culminating in a 90 to 450 kg biological laboratory in the 1977 Voyager spacecraft. Conjured up during the heyday of Apollo when unlimited budgets were projected far into the future, the ambitious Voyager program was a victim of general economic retrenchment in the late 1960s. In its stead a very small "hard lander" was briefly considered. In one design a protective balsawood shell broke open on impact, revealing a squat watermelon sized spacecraft. A camera was positioned on an extendible mast. Little else in the way of scientific equipment was included. It was recognized that the mission lacked both scientific merit and exploratory excitement. It was replaced by the more ambitious Viking which, ironically, grew to a point where it incorporated many of the capabilities originally included in Voyager.

Viking included two Orbiter Lander pairs to be placed into orbit about Mars. Following successful orbit insertion the Landers would be released and directed toward the surface. Slowed first by aerodynamic drag, then by parachute, and finally by retrorockets, they were designed for a "soft" landing. At 2.5 m sec the jolt would be something like that encountered when jumping off a 35 cm high stool on Earth. Except for the parachute phase, the entire sequence would be similar to that employed for Apollo landings on the Moon (Figure 1).

Exactly how is a decision made to fly a particular mission? NASA administrators have at their disposal a number of planning teams, staffed primarily by engineers and cost analysts. In addition, advisory committees of scientists are asked to analyze and put in sequence the various mission options. Building on this background, NASA administrators submit a specific budget with a particular mission called out by name, something termed a "line item." If the mission survives subsequent budget trimming by the Office of Management and Budget and Congress, it is elevated to an "approved" category. Various aerospace companies are invited to submit bids for the construction of the spacecraft, following the design requirements established by NASA engineers. At the same time an "Announcement of Flight Opportunity" is widely circulated among universities and research laboratories. Scientists wishing to propose a scientific experiment of their own choosing, or to participate in an experiment already slated for inclusion – a camera would be a good example – send in their credentials. A disinterested group of scientists meets to consider all applications, and then to recommend to NASA those considered best qualified.

It is a supremely democratic arrangement. Everyone can respond to the opportunity. In my own case, for several years I had been involved peripherally in mapping the Moon, using photographic information from Lunar Orbiter missions. I wanted to become more closely involved with space science, but was advised by a NASA official that there was no middle ground. Either you were a dilettante or you were an approved mission investigator. It so happened, he added, that the deadline for Viking applications was only several weeks away. Armed with little additional information, I obtained the necessary forms and started filling them out. Midway through I was tempted to chuck the whole venture. A series of questions seemed aimed specifically at revealing my inadequacies. What was my previous research on Mars? Zero. List my relevant publications. Pretty meager. List the institutional resources that would support my efforts. None. Against my better judgment I persevered, and filed the completed application.

Several months later, having heard nothing and wishing to end the whole debacle, I called NASA. To my amazement, my name was recognized, and a man told me that official announcements would be made in a few days. Conservative and skeptical though I am, I sensed that this reception hinted at good news. Sure enough, my application was approved.

The initial Lander Imaging Planning Team also included Alan Binder, an astronomer then at the IIT Research Institute; Elliott Levinthal, a physicist at Stanford University; Elliot Morris, a geologist with the U.S. Geological Survey; and Carl Sagan, an astronomer exobiologist at Cornell University. Subsequently, the team was enlarged to include Fred Huck, a research engineer at NASA Langley Research Center; Sid Liebes, a physicist at Stanford University; Jim Pollack , a physicist astronomer at NASA Ames Research Center; and Andy Young, an astronomer at Texas A&M University. We profited enormously from the counsel of Bill Patterson, Brown University, who served as team engineer, and Glenn Taylor, an engineer administrator who supervised the development of the cameras in behalf of the Langley Research Center and served as liaison between our team and the rest of the Viking Project.

My first person to person contact with Viking came when Gerry Soffen, Project Scientist, and Tom Young, Science Integration Manager, journeyed to Providence in the fall of 1968. The stated purpose was to explore a possibility that I would become leader of the Lander Imaging Planning Science Team. Basically, I suspect they were curious to meet someone they knew only by name. I recall that we had a pleasant lunch. There was excited talk about all that lay ahead. But we had no way of anticipating that it was the start of a professional alliance and personal friendship that would stretch forward, day after day, for eight years.

All early planning was conducted at the Langley Research Center in Hampton, Virginia. Our first task was to meet there, all the scientists recently selected some 60 in number and the engineers who had been considering the design of the mission for almost a year. During the meeting we heard extravagant promises regarding the scientific possibilities of Viking. It was heady fare.

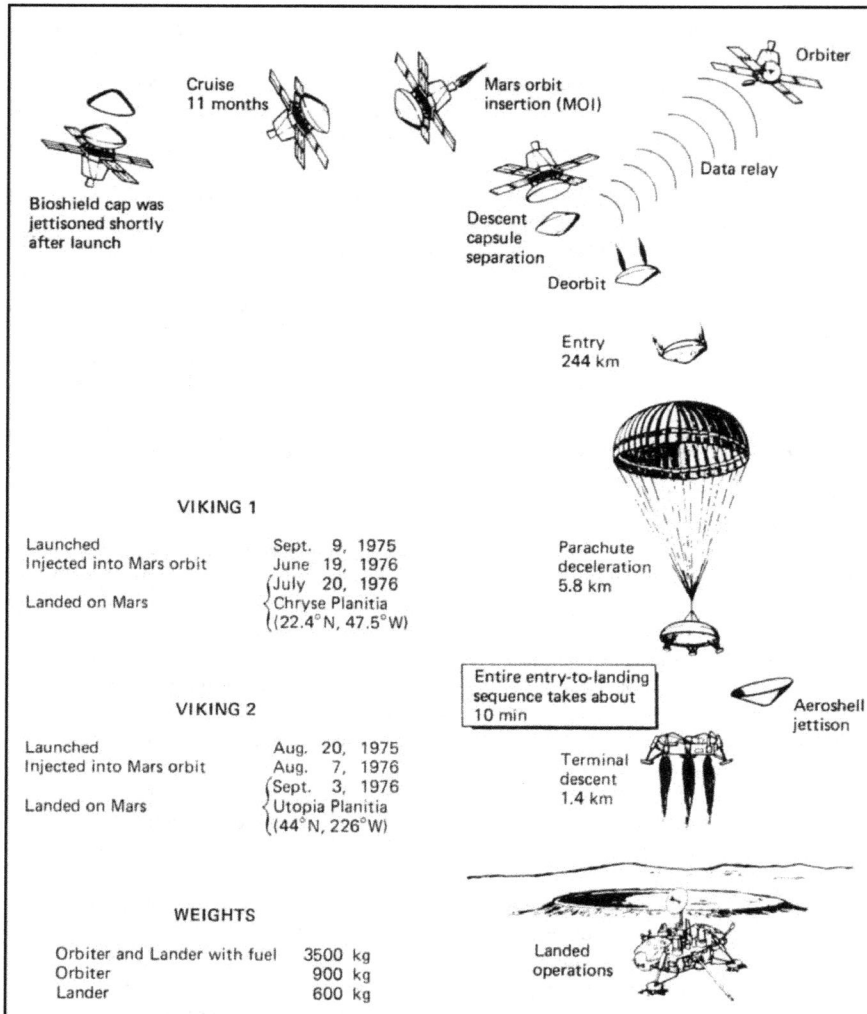

Figure 1. The sequence of events from launch to landing on Mars.

Establishing Camera Characteristics

Inevitably there is something haphazard about the initial stages of planning a complicated space probe. It is easy to talk in general terms about the scientific questions to be asked, even to list the general types of instruments to be employed. But a spacecraft is not made up of generalities. It is composed of millions of parts, each manufactured with specified characteristics to carry out a particular function.

How is the gap bridged between generality and specificity? In this instance, it is accomplished by the writing of a document that describes in engineering terms exactly what the scientist-customers wish. This document is then circulated among private industry. Any company that wishes to compete for the business makes a bid.

Note that the camera is described in "engineering" rather than "scientific" terms. There is an underlying tension separating the two. Take one of the more obvious camera characteristics, spatial resolution. The scientific goal is to take pictures of the sharpest clarity, showing the smallest detail. But you can't very well ask the manufacturer to give you a "best" picture. Instead we elected to specify that the resolution should be 0.04°. This is sometimes termed the camera's instantaneous field of view. The concept is best illustrated by looking at an enlargement of a Viking picture (Figure 2). With high magnification the image is seen to comprise a regular checkerboard of spots, each with a particular shade of gray. Each space on the checkerboard approximates a circle which subtends 0.04° as viewed from the camera position. A single trace of 360° includes 9000 checkerboard squares. A 60° swath in elevation includes 1500. A large panorama, 60° by 360°, comprises the impressively large array of 13 1/z million checkerboard squares, or pixels (picture elements).

How did we decide to specify a resolution of 0.04°? The issue was debated at numerous meetings where we were challenged to demonstrate quantitatively the increase in scientific return with increase in resolution. At one point, we were presented with two pictures of different resolution and asked to identify the one with the better resolution. The presenter was trying to develop the argument that, because we couldn't differentiate between the two, it made no difference what the actual camera resolution was. Predictably the argument collapsed when everyone correctly identified the two pictures. In the final analysis, the selected resolution of 0.04° was an educated guess of the best that was instrumentally possible.

Over the course of several years, one of our team members, Fred Huck, already had analyzed extensively the design tradeoffs between performance capability and engineering complexity. He had a thorough familiarity with engineering practicalities. It was he who sat down with Glenn Taylor, and, in a matter of a few days, conjured up the majority of detailed manufacturing specifications. Jumping ahead in our story, it is interesting to note that after nearly two Earth years of operation on the surface of Mars the cameras are still performing according to their original specifications.

Certain camera characteristics were dictated by spacecraft constraints. Notable among these were weight, power, and bit rate. The first two are obvious, but the third deserves some comment. Cameras generate a great deal of information in a very short time. This implies some mechanism for storing all that information. In a conventional camera the film is that storage device In a television camera system the data is generally

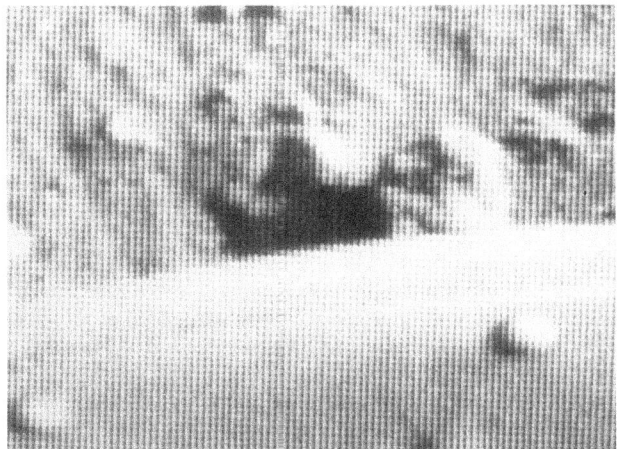

Figure 2. A part of the first picture taken on the surface of Mars, and a greatly magnified region within that picture. Note that the picture comprises a large array of discrete spots which range in brightness from white to black.

stored on magnetic tape. When the Viking mission was conceived, it was known that the entire spacecraft would have to be sterilized to avoid the possibility of carrying any Earth organisms to Mars. The procedure adopted was heat sterilization. Just before launch the entire spacecraft would be placed in an oven, and heated to 110°C for approximately 40 hours. It was feared that neither film nor magnetic recording tape would be stable at that temperature, and that surface chemicals would volatilize. What was needed was a camera that required no onboard storage device, but generated data precisely at the rate that it was being transmitted to Earth. Two transmission rates of 16000 and 250 bits per sec were available, so these same rates were selected for camera operation also.

In the spring of 1970 we met to review the six proposals for camera construction submitted by private industry. Strictly speaking the decision was not ours. The Martin Marietta Company had already been selected to assemble the entire Lander. It was their task to identify subcontractors to build the various science instruments. Their choices were subject to approval by NASA managers at Langley Research Center and Washington Headquarters. Our own team acted as a science lobby. Working outside the contractual framework, we attempted to persuade those who had to make the choices. Unfamiliar as we were with the intricacies of business arrangements, we sometimes cynically assumed that our advice would be ignored. Our concern was unfounded. As the project unfolded, our views were solicited at every point of decision. Indeed, many times our judgments were requested on subjects where our understanding was more scantily intuitive than solidly reasoned. Although flattered, I was frequently embarrassed by the willingness of managers to adopt our suggestions, while tactfully disregarding our general ignorance of spacecraft construction and operation.

As we considered the six camera proposals, one clearly ranked above the others. The technical section was crisply written. It was obvious that the proposing company, using their own funds, had made detailed preliminary calculations. The proposed camera design was elegant, avoiding failure-prone mechanically moving parts and gears in favor of electronic components. There was only one problem. The price tag was much higher than that of the low bidder. Not only that, some competing companies could point to considerable experience, much of it with NASA endorsement, in the development of cameras for planetary spacecraft. Although the Viking Project was in its infancy, concerns about escalating costs had already surfaced. We presented our case with little optimism, and were amazed when the decision was announced. The cameras would be built by the company we favored, the ITEK Corporation of Lexington, Mass.

The ITEK-built instrument is called a facsimile camera. The name is inherited from a technique in telegraphy whereby a picture is divided into a grid of small squares. The brightness of each square is converted into an electrical signal. A sequence of signals sent over the telegraph wire serves as a blueprint for registering the equivalent array of squares on photographic film at the receiving end. In this way, a "facsimile" of the original picture is produced.

In the ITEK design this general concept was refined to produce an instrument with amazing accuracy and versatility. An essential feature is that the pixels are acquired in relatively slow sequence, thereby meeting the requirement for operation without tape recorder support. It differs from a conventional camera in that at no time is a complete image recorded in the focal plane. Instead of film, there is a tiny photosensor fitted with a mask permitting it to view a solid angle of 0.04° in the object scene. For low-resolution, color, and infrared sensors the view angle is three times larger, 0.12°.

A simplified view of the camera configuration is shown in Figure 3. Light from the scene is reflected from a mirror, nodding back and forth around a horizontal axis. Focused through a lens system, it is recorded on a photosensor in the focal plane. Each time the mirror nods, light from successive points along a single apparent vertical line in the object scene is recorded by the photosensor. When this cycle is completed, the entire upper assembly of the camera moves a small amount around an azimuth rotation axis so that an adjacent vertical line is scanned. As more and more vertical lines are recorded the picture builds in the azimuthal or "horizontal" dimension, moving from left to right, and, given enough time, providing a continuous panorama up to 342.5°.

An alternate way of conceiving camera operation is to imagine you could miniaturize yourself and peer through the small pinhole in the focal plane that is the photosensor aperture. All that you would see would be a flickering light. Each change in light level would document a transition between a bright and dark region in a vertical line.

In a superficial sense the operation of the camera is remarkably simple. As it goes about its business you can watch the slotted window in front of the mirror slowly move in a clockwise arc. You can detect a regular sparkle of light as the mirror looks back at you. You can listen to the whir of the mirror, and a solid thunk as the upper housing turns in azimuth five times every second.

In fact, the camera is a complex device, involving extraordinary accuracy, reliability, and miniaturization. Several examples illustrate the point. Each pixel must be accurately positioned with relation to other pixels not only in the same vertical line, but also in adjacent lines. To assure this result the velocity of the mirror must be precisely controlled. The margin of error, the maximum allowable difference between the desired and actual positions of the upper edge of the mirror is only 0.01 mm, or one-tenth the diameter of a human hair. This accuracy has to be maintained at 512 positions as the mirror rotates through a maximum of 30° and then swings back to its starting position every fifth of a second. Vertical line-scan and azimuth-stepping servocontrol electronics compensate for a slight but significant off-axis position of sensors in the focal plane.

The camera employs an array of twelve photosensors (photodiodes), any one of which can be used to acquire a picture. Some of the diodes are mounted at different focal positions to achieve optimum focus at various distances. Others are equipped with filters. Measurement of "blue," "green," and "red" light permits construction of color images. Each line is scanned successively with each of three photodiodes to record the relative contributions of blue, green, and red light. Then the adjacent line is scanned three times. For color pictures the pixel size is 0.12° instead of 0.04°. The increase in size, with related loss of spatial resolution, is required because relatively less light passes through the filter.

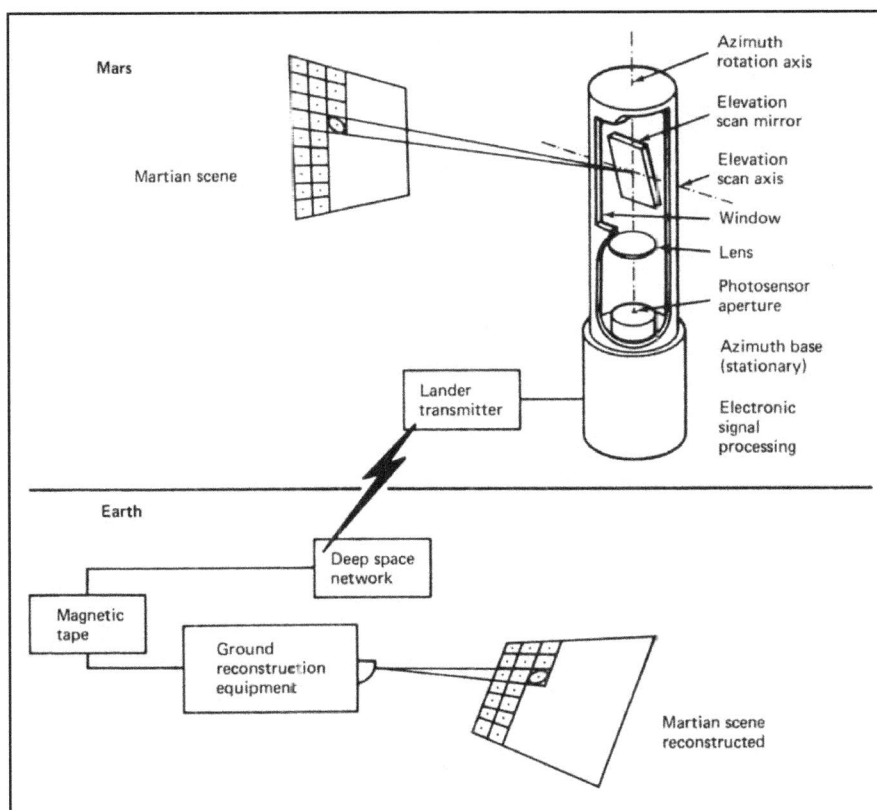

Figure 3. Schematic configuration of the Viking Lander camera and related equipment.

The photosensor array represents a difficult problem in miniaturization. The aperture over each high-resolution (0.04°) diode is 0.041 mm in diameter. Even though so small as to be barely detectable with the unaided eye, the size of the aperture must be carefully controlled, within several thousandths of a millimeter. The entire assembly, including twelve diodes and associated preamplifiers is only 3.4 cm in diameter and 2 cm high (Figure 4).

In conventional cameras exposure times are controlled by varying the lens aperture or time of exposure. In the Viking Lander camera varying exposures are attained by sampling the electrical signal in different ways. Sixty-four different signal levels, corresponding to 63 gray levels, are measured. The camera can be commanded to operate at any of six gain levels. At one extreme, the 63 sampling points are spread out across the entire dynamic range of the camera, the sort of strategy one would utilize for a scene with very high contrast and high light levels. At the other extreme, the 63 gray levels are clustered over a small fraction of the dynamic range, providing good discrimination in a scene with low light levels or low contrast. Thirty-two offsets also permit one to raise the light level corresponding to the lowest gray level without changing the increment of intensity between successive gray levels. This would be a useful adjustment in situations where contrast is low but general light level is high, and, therefore, no signals are being recorded at the lower end of the dynamic range.

Figure 4. The photosensor array, the retina and optic nerve of the Viking camera.
The entire assembly is 3.4 cm across. The 11 square circuits around the edge are preamplifiers.
The rectangular circuit at the top of the array is the command module. The 12 photodiodes are
mounted on the dark rectangular block in the middle, which measures approximately 6 mm by 2 mm.

All these operations – diode selection, sampling rates, and video signal processing – are controlled by complex electronic circuits in the lower camera assembly. As the camera design evolved, more and more circuits were crowded on the mounting boards, resulting in an impressive electronic labyrinth.

General Viking Camera Characteristics

Height	55.6 cm
Diameter	
Upper, elevation assembly	14.4 cm
Lower, azimuth assembly	25.6 cm
Weight	7.26 kg
Power	
Low-resolution imaging	34 W
High-resolution imaging	27 W
Environmental operating conditions	
Atmosphere	Earth or CO_2
Atmospheric pressure	
Earth	1000 mbar
Cruise	Vacuum
Mars	2.8 to 20 mbar
Ambient temperature	
Sterilization	+113°C
Mars	-126 to + 52°C
Hazards	up to 70 m/sec dust storm

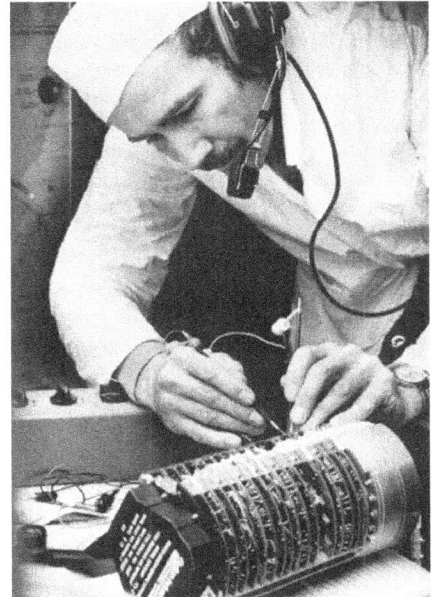

Figure 5. Checking performance of the camera electronics. The technician uses the headset to communicate with a second person who observes the quality of the video signal, displayed on a television monitor in another room.

Viking Camera Spatial Characteristics

Characteristics	Survey	Color & infrared	High resolution
Instantaneous field of view, deg	0.12	0.12	0.04
Frame width, deg.			
Elevation	61.44	61.44	20.48
Azimuth: min; max	2.5; 352.5	2.5; 342.5	2.5; 342.5
Field of view, deg			
Elevation	100°; from 40° above to 60° below horizon in 10° steps		
Azimuth	342.5°; in multiples of 2.5° steps		
Geometric depth of field, m	1.7 to infinity	1.7 to infinity	1.7 to infinity
In focus distance, m	3.7	3.7	1.9, 2.7, 4.5 and 13.3
Picture elements per line	512	512	512
Bits per picture element	6	6	6
Bits per degree azimuth	2.84 x 104	8.53 x 104	8.53 x 104
Time per degree azimuth			
Rapid scan, sec	1.84	5.52	5.52
Slow scan, min	2.0	6.0	6.0

Talking Our Way to Mars

No account of Viking would be complete without mention of the meetings. In a large program, involving many persons with different backgrounds, interests, and tasks, communication is a major activity. Reams of printed material are distributed every day. Whenever a decision of some importance was imminent – almost a daily event – a meeting was convened. There were literally hundreds of committees within Viking. In my case, the more important groups were the scientists working on the Lander Imaging Investigation, the engineers designing the camera system, and the leaders from all science teams comprising the Science Steering Group. During the eight years before launch I must have attended more than 400 meetings Initially, the opportunity to fly to some distant city was an exotic diversion. But not for long. The routes to Hampton, Denver, Los Angeles, and Orlando – localities of Viking activity – became as familiar as the quarter-mile route from my home to my office at Brown University. That peculiar disorientation in both time and space that results from a long airplane trip became an accepted state of mind. One episode stands out. I recall shuffling out of an airplane late at night after a few hours of half-sleep, and failing to recognize either where I was or to what end I was traveling. For several moments I had the Kafkaesque feeling that I had somehow lost my identity, that I had become separated from the real world.

I cannot deny the excitement of participating in this nonstop drama of crisis and decision making. Critics might question the usefulness of frantic racing around the country, with talk the only obvious product, but every meeting revealed new problems. Each person was obliged to report what progress he had been making. Cover-ups were impossible. In retrospect, thinking about all the blunt statements of disagreement and criticism, I am surprised that I can recall no instance when a participant lost his temper – at least to the point of climbing across the table and slugging his adversary. Everyone seemed to understand that the high stakes left room for neither social niceties nor aberrations.

On the positive side, helpful advice came from unexpected quarters. Useful exchanges of information prevented isolated journeys up blind alleys. When no obvious solution to a problem was apparent, we proceeded by vote. The majority opinion dictated the next step. In one sense, that appears absurd. Certain things are matters of fact. To what useful end can one vote on the proposition that a camera should cost no more than X dollars, whereas a biology instrument should cost Y dollars? Or that the average Martian atmospheric pressure is 1 percent of the Earth's atmosphere as opposed to 0.5 percent? Viewed in another context, an open meeting in which all participants have equal vote has served Americans well in many previous situations. Perhaps more than we realize, it is a method of pooling information with which we have grown up. I like to think that the ultimate success of Viking can be traced back to those countless meetings at which we chewed on one problem after another – hours of thoughtful criticism and, sometimes, clamorous sharp-edged debate.

Deadlines

Every Viking activity was framed in time. Weekly reports indicated tasks accomplished, and deadlines projected through the next few months. In conference rooms regularly updated calendars documented the days to launch, as well as hundreds of intervening events.

This emphasis on time was dictated by the nature of the journey to Mars. Approximately every two years Earth and Mars draw side by side in their respective orbits, an event known as an opposition. For a period of only a few months just before opposition, the conditions are favorable for a spacecraft to spiral out from Earth to Mars. For all other times, the thrust of the rockets is inadequate. For this reason, a so-called launch window can be identified years in advance.

Viking was first planned for a 1973 launch. During January of 1970, when the very existence of the mission was threatened by funding problems, it was decided to delay the launch two years to 1975, spreading the cost over a longer period. Our disappointment was short-lived. In fact, the delay was something of a reprieve. In retrospect, it is clear that the spacecraft could never have been designed and constructed by 1973 without seriously compromising both capability and reliability.

The ultimate deadline, then, was the August-September launch window in the summer of 1975. The spacecraft had to leave Earth at that time. No excuses. Another two-year delay until 1977 was unthinkable in terms of increased cost and administrative complexity.

Development of all instruments was keyed to the 1975 date. Working from that deadline backward, a cascading array of secondary deadlines was identified. A slip of a few weeks in 1971 could endanger the delivery of the hardware to the Cape Kennedy launch facility in 1975. Everyone understood the penalty. If the instruments were only half-ready, they would be flown half-ready-or not at all.

Working within these constraints was less of a burden than one might imagine. Indeed, it was an exhilarating change from our normal activity – in a university, at least – where the business of one day can be deferred to the next day, or even the next year. The Viking goals were sharp. There were no compromises, no rationalizations. Every problem required a timely solution.

Only when you move away from a project like Viking – and are no longer controlled by the calendar – do you realize the impact of that discipline. It affects you in small ways – always keeping an engagement calendar in your pocket, leaving meetings with just enough time to catch a late plane home and in larger ways – looking forward to a future where events yet to come assume the reality of the present. Now that Viking has passed, I sometimes feel adrift without those signposts stretching out before me through the years ahead.

Figure 6. Viking 2 was launched from Kennedy Space Center aboard a Titan Centaur 3 at 2 39 p. m. EDT, September 9, 1975. The spacecraft was placed on a trajectory that carried it into orbit about Mars in August 1976.

Design Changes

Even though the fundamental characteristics of the camera were specified in the initial contractual agreement between ITEK and Martin Marietta, some elements of the design proved either impractical or undesirable. We were continually contemplating changes.

In the first drawings submitted by Martin Marietta a single camera was shown on the Lander. It was mounted on top of an extendible vertical mast (Figure 7). In this way the field of view could be varied and stereoscopic pairs of pictures could be obtained. However, it was an unusual stereoscopic perspective, as if one of our eyes was situated directly above the other. A more compelling objection involved redundancy and reliability. Did we really want to send a one-eyed traveler to Mars? The issue was never debated at any length. Even without any supporting arguments from the scientists the project managers quickly decided in favor of two cameras.

The next issue had to do with the mounting of cameras. There were several contradictory requirements. On the one hand, we wished to have an unobstructed view of the surface and to see the distant terrain. This dictated placement of the cameras on high masts. An even more dramatic solution, briefly considered, was the installation of three cameras, one at each of the triangular corners of the Lander body. On the other hand was the requirement to conserve space and weight. A high mast added undesirable weight and, in addition, was a protuberance that could not be accommodated in the small volume between the protective covers that encapsulated the Lander in transit to Mars.

A compromise design was pursued for the better part of a year. The cameras would be mounted on hinged masts (Figure 8). During the trip to Mars they would be folded down; after landing they would be swung up, rotating through 180°. In retrospect it is difficult to see how this design concept survived for as long as it did. It introduced undesirable complexity and serious risk. During landing the sensitive camera electronics would be situated close to the base of the spacecraft, susceptible to collision with a boulder. If the swing mechanism failed to operate the cameras would remain in an inoperative position. If dust coated the exposed upper plate and base of the two mast sections, the deployed camera might be slightly tilted, introducing an unknown error in topographic analysis. Finally, if the cameras were not securely connected to the body of the spacecraft, they would not benefit from heat conducted from radioisotope energy sources. Isolated on the top of articulated masts, they might freeze to death. When all the liabilities were spelled out, the swing-up masts were discarded.

The solution was to mount two cameras on short stubby masts. (See Figures 9 to 11.) Although this fulfilled weight and thermal requirements, it dismayed the scientists. A third of the field of view below the horizon would be blocked by the Viking Lander. Especially severe was obscuration of the near field where we had our only opportunity to photograph the surface with high spatial resolution.

During a series of meetings the precise height of the masts above the surface was negotiated, the scientists reluctantly inching down and the engineers just as reluctantly moving up. We finally agreed on a goal of 1.5m. Having recently given ground on several other changes in camera design, the scientists were skeptical. Sure enough, the final height was 1.3m.

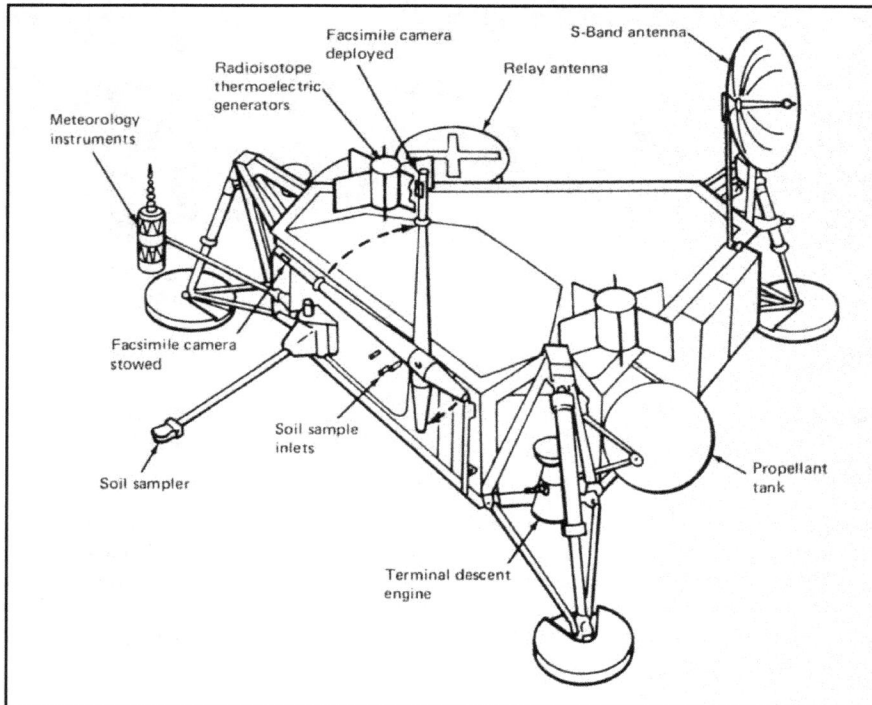

Figure 7. Initial conceptual design of the Viking Lander. A single camera is mounted on a hinged mast that can be extended vertically after it swings up.

Not everything worked to our disadvantage. Capitalizing on the conservative engineering practices utilized in the initial camera design, we were able to incorporate three additional spectral channels, something like having access to three different levels of infrared film in addition to color film for a conventional camera. As previously noted, the camera accommodated an array of twelve photosensors. Some sensors were mounted at slightly different focal positions to permit maximum sharpness at various distances. Initial analyses indicated that four focus settings were required for the 0.04° high-resolution images and that two focus settings might be required for 0.12° color images. On this basis an array of twelve sensors was specified: one for black-and-white 0.12° survey, four for black-and-white 0.04° high-resolution, six for 0.12° color, and one extra – a bonus for a symmetrical design that called for two rows of six sensors (Figure 11).

Detailed testing with prototype electronics demonstrated something we had suspected. For the large-aperture sensors one focus setting was adequate. Nothing was to be gained by placing two sets of "color" diodes at two focus positions. This left three sensors unallocated. We quickly resurrected a former request to use three filters in the near infrared, a spectral region in which we anticipated instructive absorption effects in the Martian sediment and rocks. Project managers graciously acceded to our request. In fact, it was deemed less expensive to continue with the original twelve-diode array than to redesign a smaller eight-diode array.

Someone suggested that we use the twelfth diode, the one included primarily for symmetry, for calibration with the Sun as a light source. By simply eliminating the amplifier associated with the diode, the electrical signal was reduced to a point where the Sun could be viewed directly without saturation of the electronics. Throughout camera construction the so-called Sun diode was incorporated but ignored. Nobody wanted to defend its use, but neither did it seem worthwhile to argue for its exclusion. Only when we reached Mars was the true usefulness of the Sun diode revealed. With unsuspectedly large amounts of dust suspended in the Martian atmosphere, the apparent brightness of the Sun was an important measure of the atmospheric opacity.

Some design changes arose from our concern about hazardous conditions on Mars. Though space missions appear to rely on dispassionately objective numerical calculations, subjective appraisals continue to influence the design. Before Apollo there had been concerns that the astronauts would be trapped in quicksands of dust. The same worries were resurrected for Viking. More persuasive were calculations made by Carl Sagan regarding the erosive power of wind-driven sand. His arguments were based on telescopic observations of dust storms on Mars. Given the very low atmospheric density, less than a hundredth that of Earth, the wind velocities were clearly very high, on the order of 50 to 100m/sec. Theory and experiment suggested that erosion rates under those conditions are very great, as much as one cm/yr. A twofold danger existed: first, that the spacecraft might be buried under a blanket of sand; second, that it might be sandblasted into oblivion. The horror stories were spiced by knowledge that two Soviet unmanned spacecraft had mysteriously stopped operating just as they touched down on the surface of Mars, one in 1971 and a second in 1974.

Our first concern was to protect the glass window through which the light passed on its way to the nodding mirror. The window was slightly recessed, but otherwise susceptible both to coating and pitting by dust.

Like many apparently straightforward problems, this one proved intractable – but not for lack of entertaining suggestions. A variety of mechanical brushes and wipers were proposed but rejected because of their mechanical complexity. Someone even suggested, half facetiously, that we mount a weather vane atop the camera, so that the recessed window pointed downwind when not in use. The protective strategy finally endorsed was adequate, even if lacking in elegance. Not fully believing in any one approach, we decided that safety lay in numbers. First, the window was coated with special material that resisted dust deposition and erosion. In addition, a second transparent window was installed in front of the primary window. On command, the second window could be swung out of the way. When not in use, the camera slewed to a position where the recessed window was protected behind a fixed post. During a final attack of anxiety when the cameras were almost completely built, we strapped on a device which would blow a jet of compressed air (actually carbon dioxide) against the window, thereby sweeping away any thin veneer of dust.

Other potential problems were unevenly pursued. A report that dust might adhere electrostatically stimulated a high-priority test program, but, somewhere along the line, enthusiasm for yet another protective device to dispel surface charges dissipated. Enough is enough.

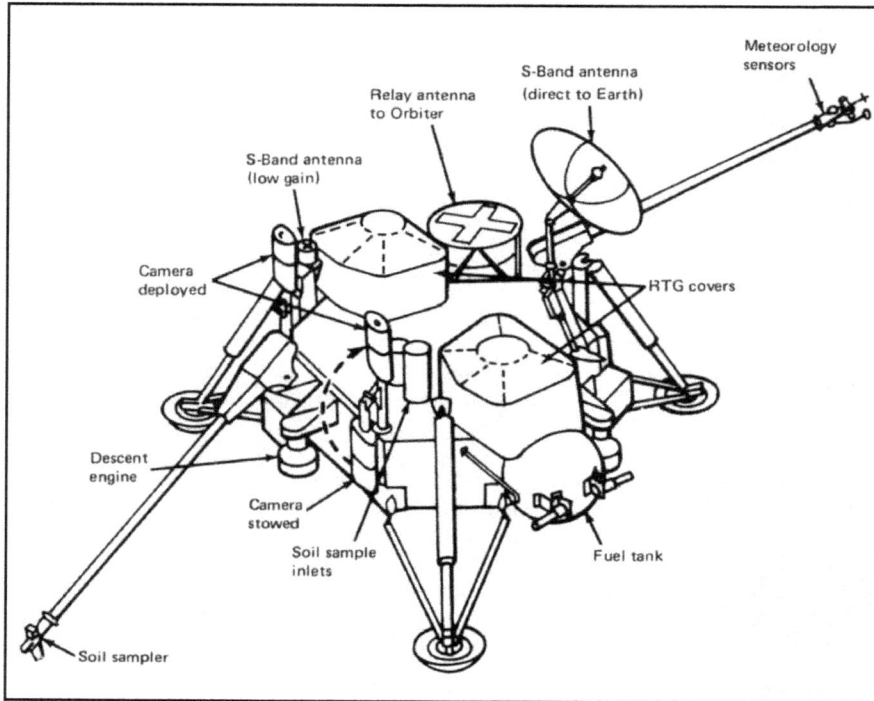

Figure 8. An intermediate design of the Viking Lander showing two cameras on hinged masts.

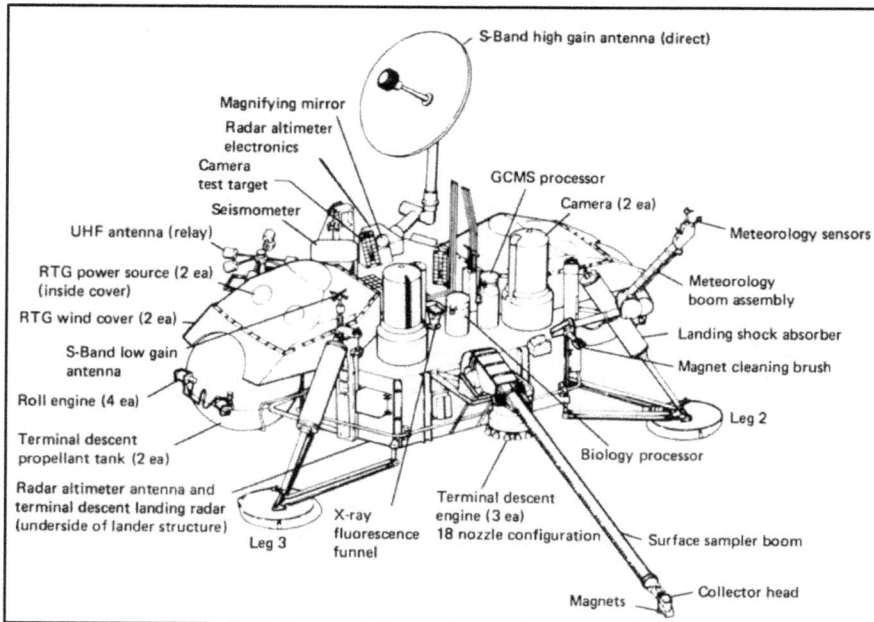

Figure 9. Schematic view of the final design for the Viking Lander.

Figure 10. The Viking Lander. Many of the science components are identifiable, including the two cameras, the meteorology boom, and the partly extended surface sampler arm. This s a spare backup Lander used for testing. During the operational mission it was used at the Jet Propulsion Laboratory to demonstrate the feasibility of surface sample sequences planned for the spacecraft on Mars. The background painting reflects the general prelanding concept of the Martian surface.

Turning our attention to other parts of the camera, we belatedly worried about the possibility of wind-driven dust seeping in and jamming moving parts. In particular, there was an external lip on the housing adjacent to the bearings that facilitated the rotational movement of the upper camera section. Could dust settling on the external lip sift through several protective seals into the bearing assembly? There was only one way to convince the fearful. One of the cameras was set up in a special wind tunnel maintained by the McDonnell-Douglas Company in St. Louis. Fine-grained rock powder was introduced, creating an awescme dust storm. Fine dust clung to every surface, but the bearings remained dust free. Only when the test continued to a point where the cameras were virtually buried did performance deteriorate. Not unexpectedly, that was due to dust caked between the protective post and the upper housing.

Manufacturing Problems

From the very first day of camera construction it seemed that nothing was built without some defect. All parts had to be fashioned from a limited list of approved materials that would withstand the deep space and Martian environments, and, in addition, would not outgas volatile organic materials that might lead to a false positive result from the biology instruments.

Batches of specially constructed electronic parts arrived from a supplier, and a check revealed that only four or five out of a hundred parts met the rigorous Viking requirements. Another batch arrived with similar results. Special supervisors from Martin Marietta and ITEK flew out to California to monitor each step in the construction of the parts. The yield increased, but rejected parts still outnumbered those that passed all qualification tests.

The elevation assembly that controlled the movement of the mirror proved an unexpected source of difficulty. The shaft was encased in ball bearings lubricated with a solid compound. After repeated mirror movements the lubricant built up at certain positions, disturbing smooth rotation of the shaft. A waiver from a general Viking rule was obtained.

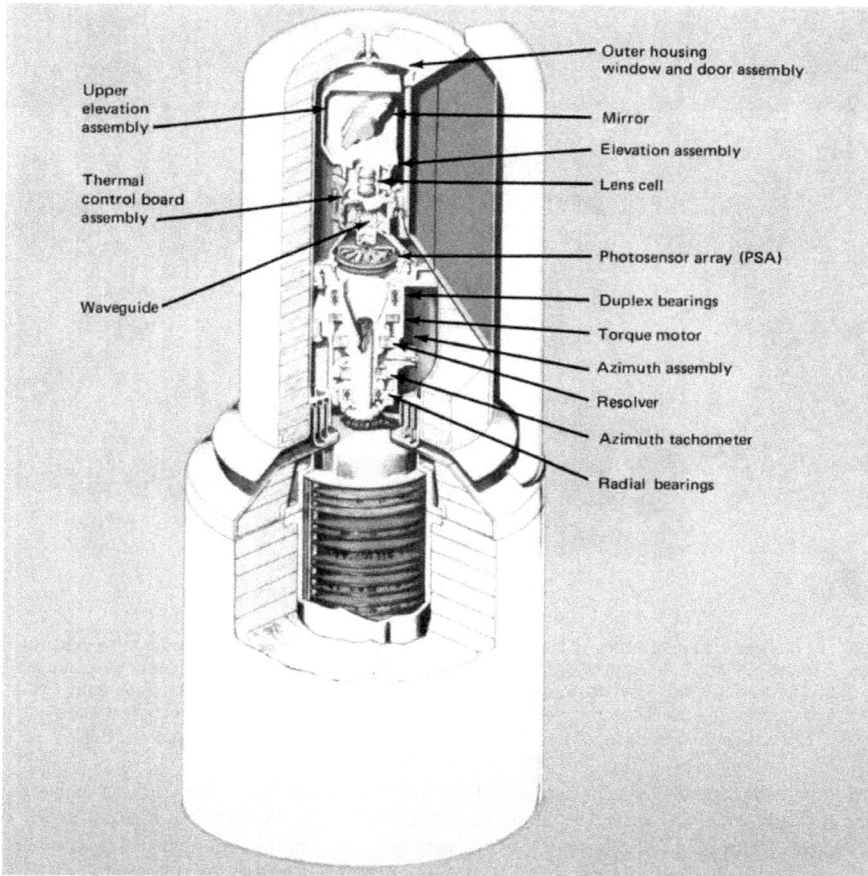

Figure 11. A cutaway drawing showing arrangement of components in the Viking Lander camera.

Because the elevation assembly was hermetically sealed, isolated from the rest of the spacecraft, a wet lubricant could be used in place of the troublesome solid lubricant.

The motor that drove the mirror was touted to be the ultimate in dependability. After long term operation, its innards were examined. The commutator brushes had essentially disappeared, ground down to nubbins. A frantic search for new brush materials was initiated. Finally a likely candidate an exotic mixture of high silver content carbon and molybdenum disulfide was discovered. A long term test showed negligible wear. Joe Fiorilla, the ITEK chief engineer, was not satisfied. Continue the testing. Can we stop now? Longer. So the motor ran on and on, far beyond the qualification requirements, a penance for all its former sins.

The most potentially devastating problem involved the tiny photosensor array (PSA). This was the single most critical component of the camera, the retina of our eye on Mars. ITEK had subcontracted the work to a company with special experience in this area of microelectronics. But as month followed month, there was nothing but bad news. Each time the sensor package was built up, parts would fail. Technicians, working under unusual tension, created inadvertent damage with a single false movement. Each miscue meant weeks of delays. To the scientists it began to appear as if the designers had overstepped the bounds of realistic possibility. In an ironic reversal of roles — it is usually the scientist who demands better instrumental capability and the engineer who adopts a conservative position — we urged the project engineers to incorporate a simpler photosensor array using only half the diodes. Fortunately our suggestion was shelved. The struggle went on, but time was running out.

At a tense meeting attended by all the chief managers of the Viking Project, an extraordinary decision was made. The contract with the ITEK supplier would be terminated, and all partly fabricated components would be sent directly to the Martin Marietta facility in Denver. There a special laboratory would be equipped to accomplish the work that had so far defied completion. Bizarre, inverted contractual relationships were forged to fit the special circumstances. NASA Langley built parts for the PSA and supplied them to Martin Marietta. That company built the PSA and furnished the units to ITEK. ITEK incorporated the PSAs into cameras and delivered them to Martin Marietta under the Viking contract let by NASA Langley.

It was a hazardous gamble. Important weeks were lost while the new Martin Marietta facility was prepared. Institutional rivalries were ignored – anyone who could help was called in. Bill Patterson, our team engineer with special background in this area of microelectronics, traveled from Brown University to Denver for a few days of consultation. Those days stretched into weeks; six months later he returned to Brown. Amazingly, by the time Bill was back at Brown, the photosensor arrays had been built. And they worked. A few months previously we would have settled for an array with one or two diodes inoperative. The components delivered to ITEK by the Martin Marietta task force were completely functional. Several units were shuttled back and forth between Denver and Boston for repair, but, at the time of final camera assembly, every diode in every assembly was ready to carry out its assigned task.

Early on, Viking managers at Langley devised a humbling technique for charting the progress of the program. The most grievous problems were assigned to the "Top Ten." At regular weekly reviews, the engineer with relevant responsibility was required to brief Jim Martin, Project Manager, on what progress had been made. More often than not, progress was backward.

Barely a year after the start of camera construction Glenn Taylor called me with the expected news – we had made the Top Ten. We tried to look on the bright side at least we wouldn't be laboring in darkness anymore. In fact, our early arrival on the Top Ten (something of a misnomer since the specially designated problems sometimes numbered up to fifteen) proved beneficial. We received helpful attention from a group of consulting engineers appropriately called the Tiger Team – before they were exhausted by the endless succession of problems that came later.

The burden of manufacturing problems was especially heavy for the several Martin Marietta engineers who were permanently in residence at ITEK, coordinating contractual and technical affairs between the two companies. On the one hand, they were the daily recipients of strident phone messages from their home institution, asking them what the hell was going on, why nothing was being delivered on time and within cost. On the other hand, the ITEK personnel were less than delighted with the intrusions of outside observers – they recognized their problems clearly enough, without having others remind them of their deficiencies.

Vince Corbett was in charge of the Martin Marietta resident group at ITEK. One Friday afternoon, as I sat in his office listening to tales of misfortune, I urged him to take a day off. Why not drive down to Providence the ITEK facilities were in nearby Boston – and spend the day sailing on Narragansett Bay? Vince accepted. He, his teenage son, and I spent a relaxing afternoon on our day sailer. It was one of those lovely crisp Indian summer days. Returning to the mooring, I made a poor approach. As the buoy drifted by to one side, Vince's son dove into the cold water to retrieve it. Somehow it seemed an appropriate end to our day of recreation – our mere association with Viking guaranteed that we would be dogged by misfortune.

One of the more vexing problems proved to have an unexpectedly simple solution. When the camera was operated in a special chamber cooled to the low temperatures prevailing on the surface of Mars the azimuthal drive jammed. When the temperature was raised the problem disappeared. Exotic electronic malfunctions were hypothesized, but common sense observations indicated the real problem. There was very little clearance between the upper camera housing and the fixed post against which the recessed window was stowed when the camera was not in use. As the temperature was lowered the post flexed and pressed against the upper housing. The clearance between post and housing was adjusted slightly and the problem never recurred.

Gradually, imperceptibly, the situation improved. The final fabrication of the cameras was accomplished virtually without incident. There were even a few moments of humor. A technician, carefully applying solder to an electrical junction, looked up to see a group of 18 visiting engineers and administrators standing around his workbench. The technician, unimpressed, remarked to the ITEK guide that it reminded him of the typical Viking philosophy – one person does the work and 18 others kibitz.

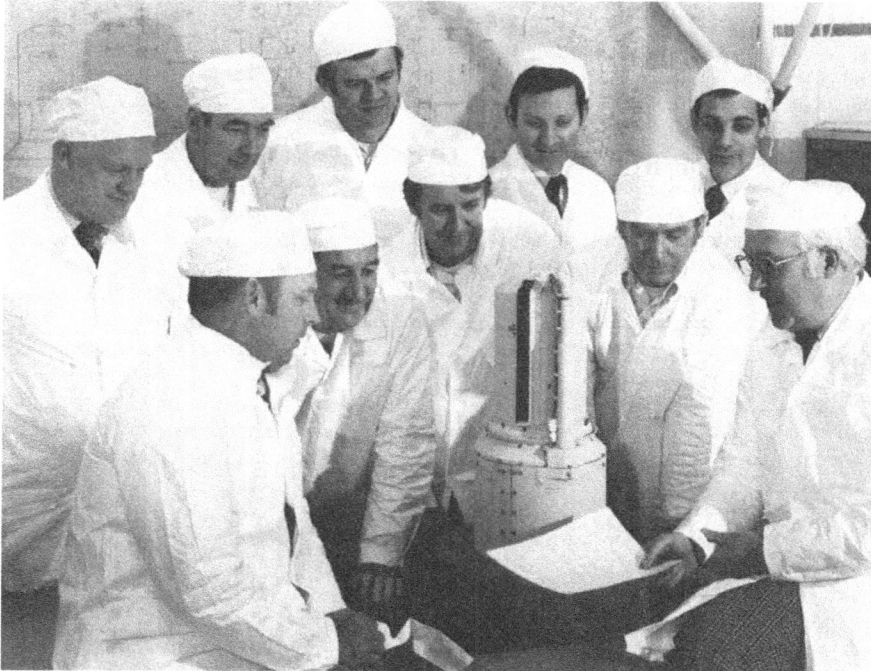

Figure 12. Members of the ITEK team dramatizing (some persons might say overdramatizing) an important event – delivery of the first camera to Martin Marietta. The upper and lower camera housings are distinguishable. The fixed post, attached to the top of the lower housing, is situated to the right of the recessed camera window.

Cameras Without Pictures

During the first years of camera development we found ourselves in the uncomfortable position of judging a complicated piece of equipment, partly assembled, that had not yet performed its primary function of taking a picture. As our anxiety increased, a difference of opinion emerged between scientists and engineers. Some, though by no means all, engineers argued that the capability of the camera could be measured quantitatively only by a series of tests involving such features as precision of pixel spacing and electrical response of photosensors. The results of the tests were generally shown as tables of figures or graphs. In terms of the contractual requirements, pictures were of little value. Only in a qualitative way did they demonstrate that the numerically defined specifications had been met.

As I pored over the dryly legal requirements of the contract under which ITEK was working, nowhere could I find reference to pictures. My engineer friends sought to reassure me. If each of the components performed according to specifications, a perfect picture must necessarily be the integrated result. I remained skeptical, partly because of my ignorance I was frustrated by schematic drawings and complicated calculations which I only dimly understood. To the hard-working engineers, already immersed in more substantial problems such as components that simply would not work, I must have seemed like the small boy who refuses to believe the Earth is round unless he can travel its complete circumference.

The more the engineers temporized, the more obdurate I became. My resolve was strengthened as others joined the chorus. Finally, at the start of one of our program review meetings, the announcement was made that a picture would be distributed at the conclusion – no doubt an inducement calculated to keep us awake through the technical reports. As that first imperfect image was passed around the table for inspection (Figure 13), the presenter began an apologetic "Let me explain . . ." The questions were sharp and numerous. What caused the shading variations? Why were some lines offset? Was this the best spatial resolution we could expect?

The same scene was destined to be repeated many times as the camera design was refined and the manufacture undertaken. For the ITEK engineers it was a choice between the frying pan and the fire. If they failed to produce a recently acquired picture, scientists and supervisors imagined the worst – the end-to-end camera system didn't work. If they did distribute a picture, then every defect was noted, generally with caustic remarks about the cost of the camera and the quality of the image.

Charlie Ross, the ITEK Program Manager, patiently tried to explain that the defects were a consequence of working with a prototype model instead of the actual flight cameras. He reminded us that the cameras were designed to take pictures on Mars, not under the uneven illumination conditions of the laboratory. His protestations sounded weak then, but we doubting Thomases appreciate now that he was right. Jim Martin, Project Manager, probably had some of those early ragged pictures in mind when he said after viewing the first pictures from the surface of Mars, that the cameras had never worked that well on Earth. Strictly speaking, of course, this was not true, but the apparent difference in quality was dramatic.

Figure 13. The first panorama taken with the Viking camera, a view of the ITEK parking lot in Lexington, Mass. The several vertical streaks, indicated by arrows, are not defects. They are the greatly distorted images of cars that drove by while the picture was being taken.

The first extensive science test of the camera was conducted in August 1971. We used a prototype camera, picturesquely referred to, in engineering parlance, as a breadboard model. To the considerable amusement of ITEK personnel, several of my graduate students arrived at the camera facility with a box, one meter square, filled with sediment and rocks. Naturally enough it was irreverently known as Mutch's sandbox. We took pictures of assemblages of sand and rocks both with the Viking camera and with a conventional film camera. To our delight we discovered that many diagnostic features were visible in Viking camera pictures (Figure 14). The images were marred by vertical banding and line mismatch, but these were problems with identified solutions.

The most important science test occurred in August 1974. By this time the camera manufacture was almost complete. Several units had been delivered to the Martin Marietta facilities in Denver where, eventually, they would be incorporated in the Lander. For several weeks we scientists were permitted the use of one of the extra cameras that, because of minor manufacturing defects, seemed least likely to be designated for the flight to Mars. (This is a continual problem for spacecraft experimenters. The best units are always carefully protected from excessive use.) After months of preliminary campaigning, we were finally granted permission to take the cameras a short distance outdoors, several hundred yards from the Martin Marietta buildings.

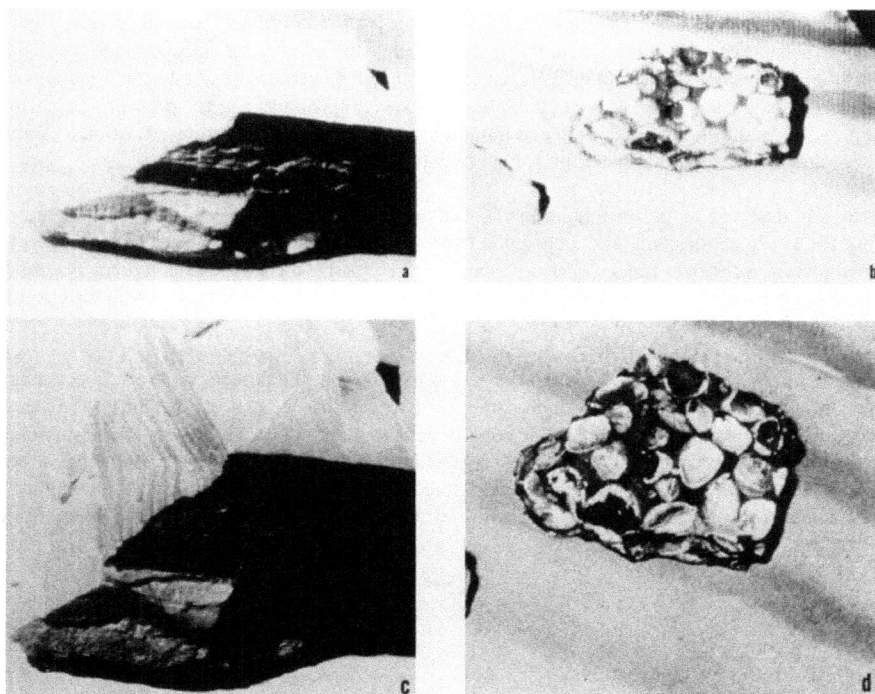

Figure 14. Pictures of (a) a layered sedimentary rock and (b) a fossil-rich rock, both taken with the Viking breadboard camera, and comparable images (c) and (d) taken with a conventional film camera.

Fortunately for the geologists among us, Martin Marietta built its plant in the scenic foothills of the Rocky Mountains where reddish sedimentary strata are tilted on edge and eroded in razorback ridges. We located an area that looked appropriately barren for a Mars analog, especially after we cleared out some dense underbrush (thereby expressing a bias about life on Mars). We spent several pleasant days at the so-called Red Rock site, frustrated only by clouds that temporarily obscured the Sun. The variations in solar intensity produced cosmetically unattractive vertical bright and dark streaking in the pictures, acquired over a duration of 10 min or more.

With the Red Rock tests completed sooner than we had anticipated, the Martin Marietta engineer in charge suggested that maybe we could venture farther afield. I was dumbfounded. Although this was precisely what we had been requesting, we had always been rebuffed with a stern lecture regarding the precious character of the cameras, and the impossibility of operating them without racks of non-portable support equipment. More than once we had peevishly asked how it could be that we were building a camera that could sustain the shock of a landing on Mars, but could not survive a short excursion on Earth.

The offer was quickly accepted. We loaded all the camera equipment and support gear in a rental truck – unaccountably, Martin Marietta seemed to have no vehicles available other than rockets – and started off for Great Sand Dunes National Monument.

To this day I have not figured out how we managed to drive blithely away with thousands of dollars worth of irreplaceable equipment. In a project where even the most trivial events were anticipated by extensive paperwork, the rules seemed to have been suspended temporarily. Several engineers accompanied the camera in the truck. The remainder of the group traveled in private cars, all of us arranging to meet the following day at Great Sand Dunes.

Bill Patterson and I eventually arrived at the monument headquarters. A government ranger confirmed that our truck was already there. From his four-wheel jeep, he pointed out the truck, halfway across a distant sand flat (Figure 15). The truck had followed a downward-sloping embankment of wet sand bordering a shallow stream. Further movement, either forward or backward, was out of the question, so we carried the camera to the crest of a nearby dune, trailing its electrical umbilical cord behind. Leaving the installation in the charge of several volunteers prepared to camp out, the rest of us repaired to a nearby motel.

The next morning, anxious to get as much work done as possible, we were up before sunrise. However, several anxious hours passed before the atmospheric humidity decreased to a level where the camera could operate without danger. (Remember that these cameras were designed to operate on Mars where the atmosphere is thin and the water vapor content very low.) Once in operation the camera and recording equipment continued to work faultlessly, better than ever before. The morning wore on and the pictures piled up.

With the obligatory images completed we turned to more frivolous projects. A box turtle and a garden snake had been obtained from a local pet store and brought along to illustrate biologic form and motion. The turtle was temporarily misplaced – one sobering but unlikely speculation was that he had been consumed by the snake, some ten times smaller – and was subsequently discovered escaping the heat under the truck.

Figure 15. Operating the Viking camera at Great Sand Dunes National Monument, Colo.
The camera is situated beneath the umbrella which protects it from the heat of the Sun.
The supporting electronic equipment is in the rental truck.

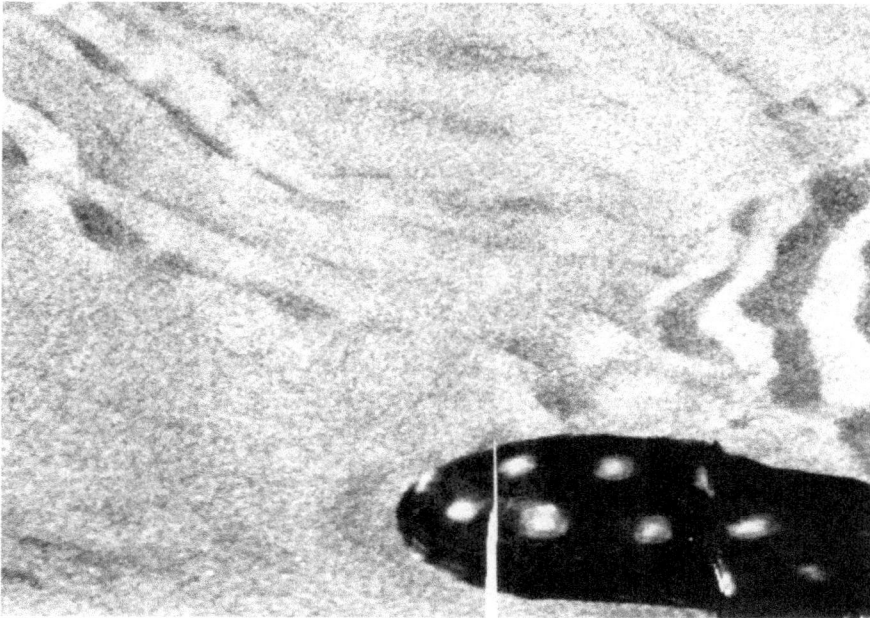

Figure 16. This picture was taken at Great Sand Dunes National Monument. A turtle about 15 cm in length,
began moving across the field of view after the camera had almost completed scanning it at rest.
Because the turtle moved in the same direction as the camera scanned, it appears more elliptical than
it really is. Turtle tracks in the background are deeper and more irregular than the natural sand ripples.

Figure 17. This picture was taken at the conclusion of testing at the Great Sand Dunes National Monument.
Because the camera scans slowly in azimuth from left to right, it was possible for some of the participants
to position themselves in the field of view several times. Artificially clipped bodies appear where
people moved before the camera had completed scanning the sector in which they stood.

We had no success photographing the snake. Every time we placed it on the hot sand it skittered away before the
camera could be turned on. With the turtle we had more luck. Several pictures recorded its artificially distended shape
as it moved across the field of view (Figure 16).

For a last picture, I suggested a group portrait. After the usual demurrer that accompanies any request to have your
picture taken, everyone lined up and the camera began its slow clockwise sweep. The opportunity for replication was
irresistible. Looking into the central slit of the camera, we could estimate when the scanning mirror had passed on to
our left. Quickly running around behind the camera we could take our place for a second portrait. In this way I
managed to be photographed a record number of seven times (Figure 17).

That evening we convened in a local restaurant for a celebration banquet. It had been perhaps the happiest day we spent on the Viking Project prior to the spacecraft's arrival at Mars. After years of ambiguous tests and reports, we had certified that the cameras really worked. Putting esoteric calculations and graphs to one side, we had said simply "I want to take a picture of that." And each time we asked – close to a hundred times – the camera faultlessly responded.

In a subtle way the success of that day's testing influenced our attitude toward the entire mission. If the cameras worked so well, perhaps it was not unreasonable to assume that other spacecraft instruments and components, plagued by manufacturing problems, might ultimately work just as well. Maybe the reams of paper outlining spacecraft performance had described reality. Maybe, two years hence, we would actually be looking at pictures from the surface of Mars.

From that point on, the testing of the cameras proceeded without incident.

Figure 18. Encapsulating the Viking Lander in its protective aeroshell. The two Lander cameras are visible.

Most of the images were used to verify that the cameras were operating according to specifications. The fabrication of the Lander proceeded, and the spacecraft was transferred to Cape Kennedy for mating with the launch vehicle. During that time the cameras were turned on thousands of times. We cautiously armored ourselves for the bad news that never came. Even following the last hurdle of heat sterilization, the cameras continued to work without fault.

The Preprogrammed Image Sequence

Viking was a project in which nothing was left to chance. At the time of launch, the computers on board the Lander were programmed to carry out a complete sequence of scientific analyses on the surface of Mars, in the unlikely event that the receivers onboard the Lander failed and we were unable to provide further instructions.

Consistent with this requirement, we programmed a series of particular pictures to be taken over a 60-sol period (a sol is a Martian day of 24 hr and 40 min). Hundreds of hours were spent in this elusive exercise how do you best arrange pictures to document a landscape that you've never seen?

We paid little attention to the preprogrammed pictures scheduled late in the mission, but the first two pictures were planned with care. In the latter case, the preprogrammed images would be the ones actually acquired. Because the first picture was initiated 25 sec after landing – and the second picture immediately after that – there would be no opportunity for a change of mind after landing. In any event, the first picture was, by definition, one that could not benefit from prior knowledge of the scene.

The planning for these first two frames was exhaustive. Everyone volunteered advice. More than a year before the landing, we were summoned to Washington to brief Dr. James Fletcher, NASA administrator, on our camera strategy. The reason for this unusual attention was obvious. In the event of a botched landing, the first two images might constitute our only pictorial record of Mars. The pictures would be transmitted to the Orbiter in the first 15 min after landing, and thence back to Earth. Not for 19 hr – including the passage of a first night on Mars – would it be possible to communicate again with the Lander.

The first preprogrammed image was a high-resolution view of the area adjacent to the footpad, the second a low-resolution panorama covering most of the viewing area accessible to the one camera (Figure 19). A number of our colleagues challenged the priorities – "If you were transported to an unknown terrain, would you first look down at your feet?" Indeed, in a common mental image, the explorer shades his eyes, looking far away to the distant horizon. Our counter argument was relatively pedestrian. A primary photogeologic goal, perhaps because it is so easily

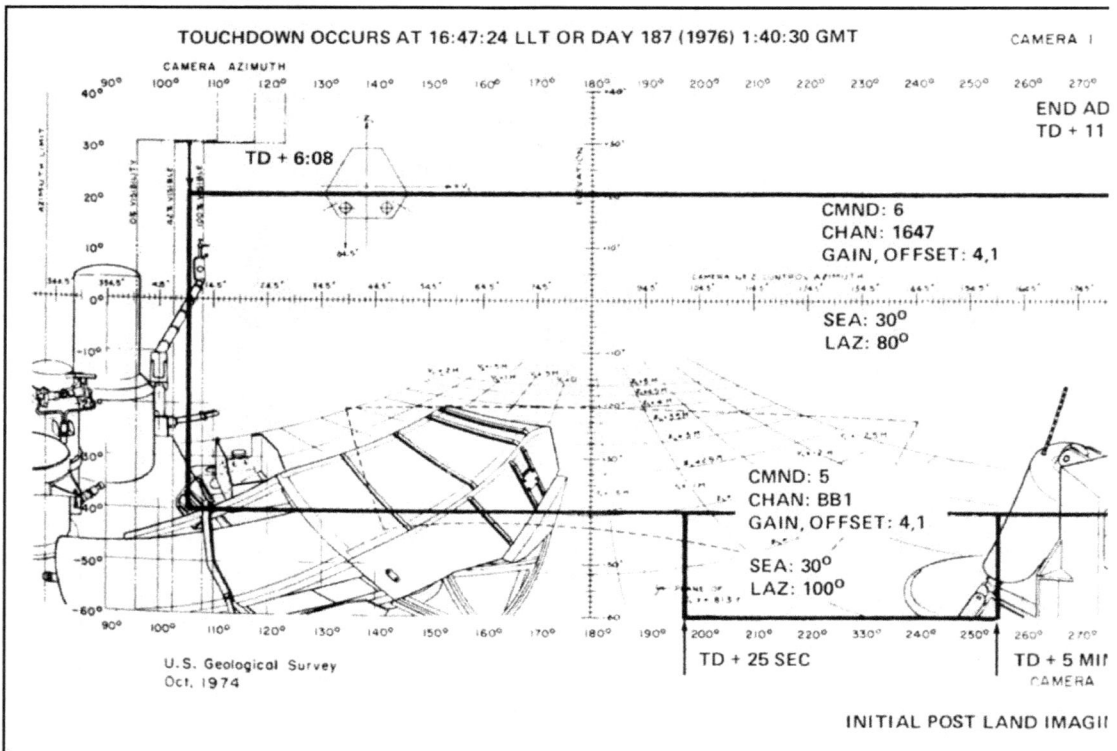

Figure 19. Outlines of the two pictures preprogrammed to be taken within the first 15 min after touchdown on Mars.
Compare the camera's "skyline" view of the Lander with the perspective drawings of the spacecraft in Figure 9.
The times listed in the upper left correspond to the originally scheduled July 4 landing.

quantifiable, is increase in linear resolution. Looking nearly straight down, the slant range was about 2 m, yielding a linear resolution of approximately 2 or 3 mm. Looking toward the horizon, nominally 3 km distant, the linear resolution would be reduced by three orders of magnitude.

Our logic would have been persuasive if the surface of Mars had been generally flat, but covered with small objects of unusual form. As it turned out, this was not the case. The rock-littered surface in the near field is relatively undistinguished, but the undulating topography and diverse geology of the middle and far field is spectacular. From both an exploratory and scientific perspective, the panorama to the horizon is the more impressive of the first two pictures.

After the Launch

Following two successful launches, the first on August 20, 1975, and the second on September 9, 1975, we looked forward to a vacation. It seemed that the hard work was over, at least for the next few months. The cameras were on their way to Mars; we were freed from any more hardware decisions. Commands for the first pictures were preprogrammed in the Lander computers. What more could we do than sit back and wait during the 10-month journey to Mars? We soon found out.

The Viking plan called for a mission operations team of more than 800 persons to be in residence at the Jet Propulsion Laboratory in Pasadena, Calif., directing the spacecraft once they arrived at Mars. The majority of these 800 persons would be visitors at JPL. Large groups arrived from Martin Marietta and from Langley. Representatives from many subcontracting companies were on hand to monitor the performance of particular instruments. The scientists, of course, came from many institutions throughout the country.

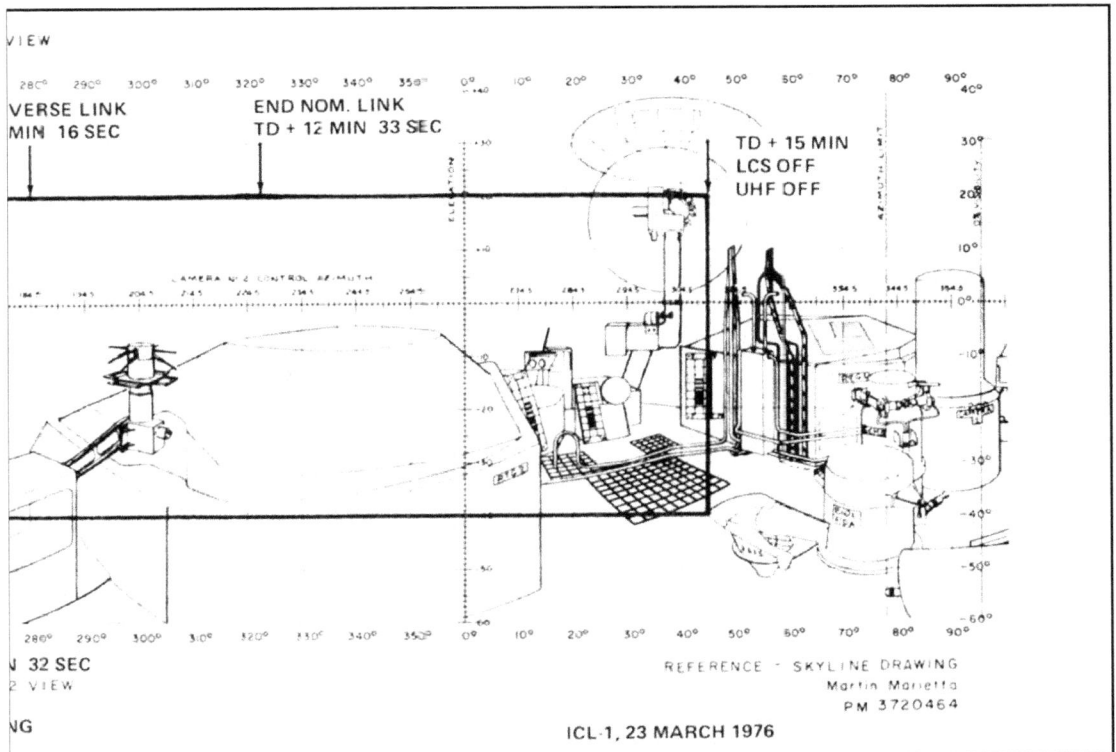

Figure 19 (continued)

It was obvious that there were problems, technical and social, in assuring that this large group worked together effectively. If everyone dropped in at JPL just a few days before the landing, chaos would result.

The proposed solution was yet another series of tests. All of us were encouraged to be in residence at JPL, starting in January 1976. I greeted the news with enthusiasm. It seemed a delightful way to escape a New England winter, and surely there would be ample opportunity for travel and relaxation in southern California. In the months that followed my wife pointedly reminded me of my prophetic gifts. Between January and November 1976, we had time for just two weekend trips.

Our first task was strictly bureaucratic. People from many backgrounds had to be blended together in functioning groups. There were the usual office politics – assignment of office space, secretarial services, and the like. Other problems were more intriguing. Scientists – especially those working in a university environment – are notoriously independent. They had no objection to being under the nominal control of group leaders from JPL and Martin Marietta, but real control was another matter. With their investigations in the balance, they rebelled against the occasional group leader who attempted to keep all decisions to himself. After a few weeks of job shuffling and personality testing, we finally settled down to the job at hand.

A series of tests was designed to simulate the actual mission operations. To provide a sense of reality, a group of "instructors," working in secrecy, designed a series of mission events that were programmed into the computers. Accordingly, the computer outputs simulated transmissions from a spacecraft on Mars.

The first test, three days in duration, assumed that things were working well. Our only job was to prepare updated command loads to respond to specific situations on Mars. We discovered early a frustrating problem that was to plague us throughout the mission. Any change in the sequence already stored in the Lander computer – even an apparently minor one – required days, if not weeks, to implement. The reasons were several. First, decision making involved a pyramidal structure. All decisions initiated at the bottom of the pyramid required approval at higher tiers, in many cases all the way up to Jim Martin, the Project Manager. Every decision required a meeting. The meetings were seemingly endless, sometimes even overlapping.

A second problem involved safety. To a degree, operation of any one instrument on the spacecraft jeopardized the operation of another instrument. Simultaneous operation might result in a power drain or a computer failure. Less dramatic perils lurked in the background. Designation of a picture somewhat larger than originally planned might utilize all the recorder tape, leaving no room, for example, for some previously planned biology data.

This maze of interconnected events required an incremental approach. A week or so before a command update was scheduled, a candidate sequence of events was determined. Then it was run through a preliminary computer program to detect errors. Inevitably some were found, corrected, and the computer's approval again requested. Eventually passing this hurdle, the sequence of events was translated into the special command words that would be transmitted from giant antennae on Earth to the Lander on Mars. Each final sequence was double checked by using it to operate a test vehicle on Earth, guaranteeing that the commands produced the desired result.

Clearly, nothing like real-time control of the spacecraft on Mars was possible. Even if there were no other constraint, the vast Earth-Mars distance meant that it took about 20 min for a radio signal from Earth, traveling at the speed of light, to reach Mars.

The tests became more challenging when anomalies were introduced. Our instructors would simulate some disaster, leaving it to us to identify and respond to the problem. If our diagnosis was faulty, our corrective surgery clumsy, we were required to repeat the test. It was like a recurring bad dream come to life. We were back in school, suffering through final exams.

One anomaly test, simulating the period just before landing, was particularly disconcerting. The problem was straightforward, but project engineers seemed unwilling to acknowledge transparent truth. The cameras were scheduled to be turned on, in this way confirming that they were still working after their long trip to Mars. There were two tests to be conducted. First, a small light was turned on inside the camera which flooded all the photo diodes. Although no image could be acquired in this way, the electrical signal recorded by each diode indicated that it was operational. In the second test, a small light bulb installed on the protective post covering the recessed camera window was turned on. In this so-called scan-verification test a conventional, if somewhat drab, image of the circular light was obtained. A single diode was assigned to the scan-verification sequence; other diodes could not be substituted.

In the anomaly test simulating the prelanding checkout, the flooding of all the diodes with the internal light indicated that one of the 12 was not working. That one inoperative diode was the same one used for the scan-verification test. Predictably the scan-verification test produced no data. The solution appeared simple. All we had to do was avoid using the one faulty diode. But, we were asked, how could we guarantee that there was not a double failure? How could we demonstrate that the scan motors had not failed along with the one diode? How could we demonstrate that further operation of the cameras would not endanger the entire Lander? The more we protested, the more complicated became the solutions. Mercifully, the test was

Figure 20. The small object to the right of the larger rock is a fossil trilobite, photographed during one of the mission tests several months before the Mars landing.

ended by a phoned-in bomb threat (not part of the planned simulation). By the time the buildings were evacuated, searched, and normal schedules reestablished, no one could recall just what had happened during the final critical moments of the simulation, when a decision on landing should have been reached. We were depressed. Surely, we told one another, nothing in real life could be as bizarre as these simulated concoctions.

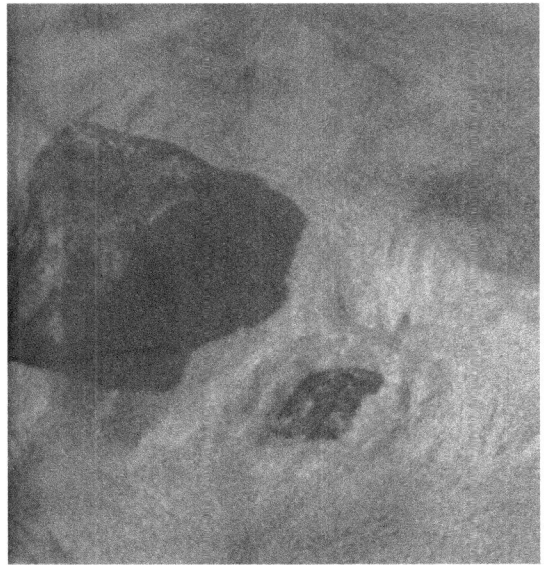

One of the final tests was a pleasant change from our previous problems, regardless of the fact that they were simulated. The landing sequence was reviewed and we sat in front of the television consoles as the first pictures from "Mars" were displayed. Of course, these pictures had been taken at a nearby locality and were appropriately coded to simulate a transmitted picture from the Lander. As we watched the first image appear, an undistinguished sandy surface was revealed. But then, in the midst of all this sand, a symmetrically jointed lobate object, perhaps 15 cm long, appeared (Figure 20). Alan Binder, standing next to me, exclaimed, "That's a trilobite!" (Trilobites are primitive arthropods that thrived in the early Paleozoic seas on Earth, hundreds of millions of years ago. Their external skeleton, with head, body, and tail, creates a distinctive fossil that is much prized by professional paleontologists and amateur rock hounds.)

Without examining the Viking pictures in any detail, we jumped to the conclusion that some lighthearted geologist had salted our Martian scene with a terrestrial fossil. At the "press conference" held a few hours later – yes, even the press conferences were simulated – we reported, tongue-in-cheek, that life had been discovered on Mars. Not everyone was amused. Some of the biologists were irritated by our hasty pronouncements. Even though we were playacting, biology was a controversial subject. Other scientists looked at the picture and doubted our identification. The more I studied the elliptical object, the less confident I became. It was clear that our initial identification was intuitive. But what if our intuition was wrong? It would not speak well for our credibility during the actual mission.

With the test completed, we were allowed to see the scene that had been photographed. Our fears were unfounded. There, nestled among the sand ripples, was a magnificent replica of a Cambrian trilobite.

The Viking 1 Landing

On June 19, 1976, Viking 1 successfully achieved orbit about Mars. There followed an exhausting month while pictures taken from orbit were analyzed and the search for a safe landing went on – and on. The landing had been planned originally for July 4, but the predesignated site was judged to be too hazardous. Alternate sites were identified, only to be revealed as equally hazardous when more pictures were acquired from orbit.

Marathon meetings served primarily to underline our ignorance regarding the actual hazards of any site. Two types of evidence, neither very persuasive, were considered. The pictures from orbit had best resolution on the order of 50 m, considerably greater than the size of objects that might effectively destroy the Lander. Radar signals from Earth, bounced off Mars and returned to Earth, were influenced by the roughness of the surface at the scale of a few centimeters, but the exact character of that influence was arguable. Perversely, all areas that were acceptably smooth in pictures appeared rough to radar, and vice versa. It seemed as if we were doomed to some sort of orbital purgatory.

Finally, the Project Manager was persuaded that an acceptably smooth site existed at 22°N, 48°W in Chryse Planitia, the Plains of Gold. The landing was scheduled for July 20. An answer to the unspoken question that had been in our minds for eight years was close at hand.

Thirty hours before Orbiter-Lander separation the two cameras were turned on for the first time in nine months, a rerun of the simulated event which had created such confusion a few weeks previously. This time everything worked perfectly. The numbers telemetered back to Earth and disgorged on stacks of computer printouts indicated that the cameras were operating just as we had anticipated. Strangely, as I looked at those pages of stark numbers, I felt the sort of joy that passes between friends long separated and once again united. Our companions, the cameras, were alive and well.

Figure 21. Steve Wall and Ken Jones (with the helmet) several hours after the successful landing of Viking 1. Wall and Jones were part of the FOVLIP team that reconstructed the first pictures as the data were radioed back from Mars. (Photograph by Richard E. D'Alli)

Separation occurred at 1:51 a.m., Pacific Daylight Time, July 20. Slowly at first, and then more rapidly, the flight path of the Lander branched from the Orbiter, bending down toward Mars. The final descent through 6.1 km took less than two min. At 5.9 km a parachute was deployed. Two sec later it was fully inflated. Seven sec later the aeroshell was ejected and the landing legs deployed. Fifty-one sec later the terminal descent engines ignited. Two sec later the parachute and base cover were separated. Forty-four sec later the Viking 1 Lander touched down on the surface of Mars.

It is not my intention to review here all the events and scientific results following that first landing. The camera results are best appreciated by leafing through the folio of pictures that makes up the major part of this volume. With the exception of the seismometer on Viking 1, all instruments on both Landers operated as expected. A wealth of information concerning the Martian environment is now in hand.

Predictably, interest has focused on the results of the biology investigations. Since all three of the experiments designed to test metabolic activity of a microbiota yielded "positive" results, it is tempting to conclude that life exists on Mars. Indeed, some of the early meetings, in which preliminary biology results were reported, were charged with the excitement of profoundly important speculations about to become historical reality. We now recognize that the biology results can be explained by inorganic surface reactions in the absence of any living forms. Strengthening this conclusion is the absence of organic compounds, documented by a gas chromatograph mass spectrometer (GCMS) experiment. But surely this does not prove the absence of life on Mars, only its absence at two localities purposely chosen to be bland and featureless. It remains possible perhaps unlikely, although statistics in this instance have little validity – that life exists elsewhere on Mars in some special environmental niche – or that it existed millions of years ago.

The First Color Picture

During the first day following the Viking 1 early morning landing we were preoccupied with analysis and release of the first two pictures which, in quality and content, had greatly exceeded our expectations. Almost 24 hr later, when the Orbiter once more passed over the Lander, another picture was relayed to Earth, this one in color.

In contrast to the attention we lavished on planning of the first two black and white pictures, we were dismally unprepared to reconstruct and analyze the first color picture. In a general way we understood that thorough preflight calibration of the camera's spectral sensitivity was mandatory. We also knew that we would need software programs that efficiently transformed the raw data. What we failed to appreciate were the many subtle problems which, uncorrected, could produce major changes in color. Furthermore, we had no intimation of the immediate and widespread public interest in the first color products – for example, intuitively corrected color images were shown on television within 30 min following receipt of data on Earth. Although we struggled to delay the deadline, we were obliged to release the first color prints within 8 hr after receipt of data.

As previously mentioned, there are three sensors with blue, green, and red filters in the focal plane of the camera. These record the radiance of the scene in blue, green, and red light. However, the multilayer interference filters used in the cameras (simpler absorptive emulsion layers would have been degraded by preflight heat sterilization) have very irregular spectral response. The blue channel, for example, responds slightly but significantly to infrared light. The extraneous parts of the signal must be subtracted, so that the absolute radiances at three specific wavelengths in the blue, green, and red are represented. A color print is produced by exposing conventional color film to separately modulated beams of blue, green, and red laser light, scanning the film with the same geometry employed in the camera.

Preflight calibration of the camera thoroughly characterized the sensitivity of each sensor filter combination. Qualitative tests indicated that simple normalization of the voltages for the three color channels – disregarding spectra leaks and displacements – was sufficient to produce reasonable color images. In making that judgment our attention was generally directed to saturated colors in the natural scene and test target.

When the first color data from Mars were received on Earth, we immediately used the same normalization techniques to calibrate the image. The result was surprising and disquieting. The entire scene, ground and atmosphere alike, was bathed in a reddish glow. Unwilling to commit ourselves publicly to this provocative display, we adjusted the parameters in the calibration program until the sky came out a neutral gray. At the same time, rocks and soil showed good contrast; the colors seemed reasonable. This was the picture released eight hours after receipt of the data. But to our chagrin the sky took on a bluish hue during reconstruction and photoreproduction. The media representatives were delighted with the Earth like colors of the scene.

Meanwhile, continued analysis supported the reality of an orangish tint throughout the scene, the atmospheric color resulting from small suspended soil particles. Several days after the first release, we distributed a second version, this time with the sky reddish. Predictably, newspaper headlines of "Martian sky turns from blue to red" were followed by accounts of scientific fallibility. We smiled painfully when reporters asked us if the sky would turn green in a subsequent version.

Our work over the past year has demonstrated that, even though the sky will not turn green, there will be nevertheless a long series of color images, each better than – or at least different from – its predecessor. The initial images were unnaturally saturated, or rich in chrome. Reducing the saturation makes the scene appear more drab. The reader should keep in mind that color is defined according to three properties: hue, the characteristic wavelength of the pigmenting material; saturation, the amount of pigmenting material; and brightness, the admixture of pigmenting material and gray background. Most persons equate apparent color with hue. However, changes in brightness and saturation can produce images which appear to be much different in hue.

More important than correcting for saturation was accounting for spectral "out of band" filter transmittance. In order to do this it is necessary to examine, for each picture element, reflectances in all six bands, three in visual color and three in near infrared. The amount of off axis radiance for each channel is estimated by considering the radiance in the five other channels and developing an integrated spectral solution that satisfactorily accounts for the measured reflectances in all six channels.

The reader may be thinking that problems of color reconstruction largely could have been circumvented by mounting on the spacecraft targets of known color. In fact, such an array of color chips was mounted and has been imaged. We have discovered that accurate color reconstruction of this target is necessary, but not sufficient for certification of good color in the natural scene. Errors are not particularly noticeable for saturated colors but become significant at low values of brightness and saturation.

In summary, the color of the Martian scene, perceived by the necessarily abnormal eyes of Viking, is elusive. In response to the inevitable question: "Is that exactly how it would look if I were standing on Mars?" a qualified "yes" is in order.

Uplink and Downlink

After the excitement of the first days, our work fell into a regular pattern. Each of the more than 40 persons on the Imaging Team was specially trained to carry out a specific task. All those months of tests began to pay dividends.

The principal division of work was between "uplink," preparing commands for new pictures to be telemetered to the Lander, and "downlink," processing the pictures that had been returned to Earth.

The uplinkers had the most demanding and tedious job. A new command load had to be prepared and certified approximately every two days. In one sense, the requirements for change in our preprogrammed sequence were less than we had imagined. The cameras were working perfectly; there was no need to switch pictures from a faulty photosensor. Indeed, even the estimated exposure settings, the so called gains and offsets, were generally correct. The entire Martian scene was of equal interest so the strategy of complete coverage displayed in the preprogrammed sequence was appropriate for the actual mission.

Most changes in pictures were dictated not by us, but requested by others. Scientists and engineers alike were reluctant to conduct spacecraft operations without a picture to document the event. For example, on sol 2, the sampler arm jammed in an intermediate position because the arm had not extended far enough to release a locking pin. Although the corrective procedure was relatively straightforward, it was decided that no further commands would be issued until we acquired a picture of the sampler arm in its stalled position. A last minute camera sequence was prepared; the resulting picture documented the suspected failure mode.

Throughout the mission, sampler arms on both Landers proved to be prima donnas. Several times they unexpectedly stalled in the middle of sample acquisition sequences. Each time, a disabling permanent failure was feared – a disastrous event since the key scientific instruments inside the Lander depended on the sampler arm to provide Martian soil. However, a new sequence of commands always brought the sampler arms back to life. This oft repeated sequence of failure despair success elation imparted a Hollywood movie like aura to the mission – it also heightened the impression of an "intelligent" robot on Mars.

The requirement to monitor the progress of the sampler arm with repeated pictures had some peripheral benefits. Our ability to generate commands for new picture sequences on very short notice was verified. As some of us had expected, the burdensome series of meetings and time consuming cross checks could be abbreviated. With the feasibility of late commands verified by these Lander health pictures, it was relatively easy to persuade managers that the same techniques could be used to update science sequences.

Pictures were required not only to certify the condition of the Lander; they were also necessary to document each sample sequence. First the candidate area had to be stereoscopically imaged to locate rocks and slopes that might endanger the sampler arm. Special computer driven equipment, constructed by one of our Science Team members, Sid

Liebes, was used for rapid construction of topographic profiles. Then a series of pictures was required during the sampling operation to document each event: successful deployment of the arm, digging of the hole, delivery of the sample to the inlet ports on the upper deck of the spacecraft. These were extraordinarily difficult sequences to plan since the pictures had to be precisely timed to intersperse with the sampler operations.

The downlink activity was equally demanding. Each time a transmission occurred from Lander to Earth, either directly or through the Orbiter relay, a team of two or three persons monitored the incoming data, made preliminary adjustments, and transferred the data to magnetic tape, suitable for further processing. As each picture was assembled on the screen you were the first to view this part of Mars. It took no unusual flight of fancy to imagine yourself on the Martian surface. The effect was heightened during those downlinks late at night or early in the morning when the flight operations building was deserted, with only a skeleton crew to monitor spacecraft operation and to catalog the data returning to Earth.

Once the raw data had been received in the FOVLIP area (i.e., First Order Viking Lander Imaging Processing – Viking acronyms numbered in the thousands, and some conversations seemed to be in a foreign tongue), it was passed on to the Image Processing Laboratory where the pictures were enhanced to produce the best possible image. Although this procedure appears questionable, it is in fact entirely appropriate. No detail is introduced into the scene that was not, in fact, recorded by the cameras. Since the raw imaging data were generally of high quality, enhancement was usually limited to restoring lines of data that had been misplaced during transmission, improving picture contrast, and adding labels and marginal scales.

Superimposed on the uplink and downlink tasks was a constant succession of science meetings and press conferences. Scientific data were coming in so rapidly that keeping abreast of each new finding was a hopeless task. We had little time even to study the pictures we had taken. No sooner was one sequence planned and acquired, than we were in the midst of the next one. Although the frenzied activity of the first few days subsided, the operation of the mission continued with around the clock activity that consumed all our resources.

The Viking 2 Landing

Viking 2 went into orbit about Mars on August 7, 1976. Searching for an environment most favorable for a Martian biota, a decision was made to go to higher latitudes where small amounts of H_2O ice might temporarily exist in or below the surface. Once in orbit we were required to land in the vicinity of 48°N.

The entire latitude band was searched. None of the pictures was very reassuring. Concern was heightened by the awareness that at the first landing site there were several rocks in the vicinity of the Lander that could easily have destroyed it, had the touchdown point been shifted a few meters. The Project Manager decreed that the second site should be rock free – or, at least, that persuasive arguments in that direction were mandatory. As in the case of the first landing, the juggling of potential landing sites was both exhaustive and exhausting. As the landing was delayed, the optional strategies for site certification and simultaneous operation of four spacecraft became Byzantine in complexity. The final decision to land on September 3 was forced less by objective knowledge that we had found a safe site than by knowledge that the flight team was running out of steam and time.

Viking 2 was targeted to 48°N, 226°W, a spot in Utopia Planitia. The name, at least, augured success. However, an undefinable pessimism preceded the second landing, strange in light of the success of Viking 1. We realized, perhaps, how quickly the euphoria of past months could change to criticism and disappointment if the second half of the total mission were unsuccessful.

The Viking 2 Lander separated from its Orbiter at 12:40 p.m. on September 3. Minutes later, telemetered data from the Lander to Orbiter abruptly stopped. The cause was uncertain, but the prospects looked dim. It was difficult to imagine that a serious communication failure essentially at the time of separation was not the expression of a more fundamental problem. Had the Lander incorrectly separated? Was it perhaps still dangling from the Orbiter? Was it hurtling toward Mars, out of control?

Within the hour it was demonstrated that the Orbiter had rolled so that its high gain antenna pointed away from Earth. It was suspected that the loss of control was due to the failure of an inertial reference power source. Fortunately there was a redundant power source. By switching to the backup the Orbiter was reoriented, but not until several hours after landing.

Meanwhile, we could track the descent of Viking 2 by using a low gain antenna on the Orbiter to monitor the signals sent by the Lander. Although little in the way of detailed data was received, the very presence of the Lander signal meant that it was operational. During an expected communication blackout immediately preceding landing, we listened as the mission controller called out "Touchdown minus seven, six, five, four, three, two, one, zero." There was a long silence. It seemed like minutes. Over the loudspeaker came a muffled prayer, "Come on, baby." Nothing. I looked down at my shoes. I remember thinking, "You always wondered what a failure would feel like. Now you know." I mentally composed some remarks for friends standing nearby. Finally, the silence was broken. "We have touchdown." Pandemonium broke loose a completely unrestrained celebration unlike the more self conscious moments following the Viking 1 landing. For a few hours, with regular communication from the Lander still not established, there was nothing much to be done. Cases of champagne magically appeared. The tight security procedures were temporarily suspended. Families trooped into the laboratories to share in the excitement. Two successful landings! It was an outcome that not even the most optimistic had looked forward to.

It was not until late that evening that regular communication with the Orbiter could be established. Now it could relay the pictures that, almost eight hours previously, had been transmitted from Lander to Orbiter just after touchdown. We were already exhausted after a long day. The event held little of the drama associated with the first Viking 1 pictures. Indeed, at first glance, the Viking 2 landscape was something of a disappointment, looking superficially like the Viking 1 scene. The profusion of large blocks was the source of wry amusement. Jim Martin reminded us that we had promised a landing on undulating desert dunes.

Figure 22. Site certification meeting for Viking 2 landing. (Photograph by Hans Peter Biemann)

In the subsequent weeks I revised my opinion of the Viking 2 site. It is, in fact, quite different from the first site, and the geological histories are probably dramatically diverse. Nonetheless, as the days passed by, I could feel my enthusiasm waning, the Viking experience slipping away. I looked forward to the end.

In early November the primary mission concluded. Mars passed behind the Sun and communication with the four spacecraft was temporarily impossible.

We crossed off the last few days on the calendar, just like the last days of school. The very last night I stayed late at the laboratory, writing notes to a few of the people with whom I had worked so closely for so long. The next morning my wife, youngest daughter, and I started off – a roundabout route back to Providence by way of the Sierras. We spent four days in Yosemite. The valley was virtually deserted this late in the season, the ground thickly covered with crisp aspen leaves. The Sun was rich and warm, refusing to give way to the scheduled appearance of winter. At night we could look out of our tent, across a field, to the cliffs bathed in the chalky, white light of a full Moon. When I think about Viking, my thoughts often return to that Yosemite epilogue.

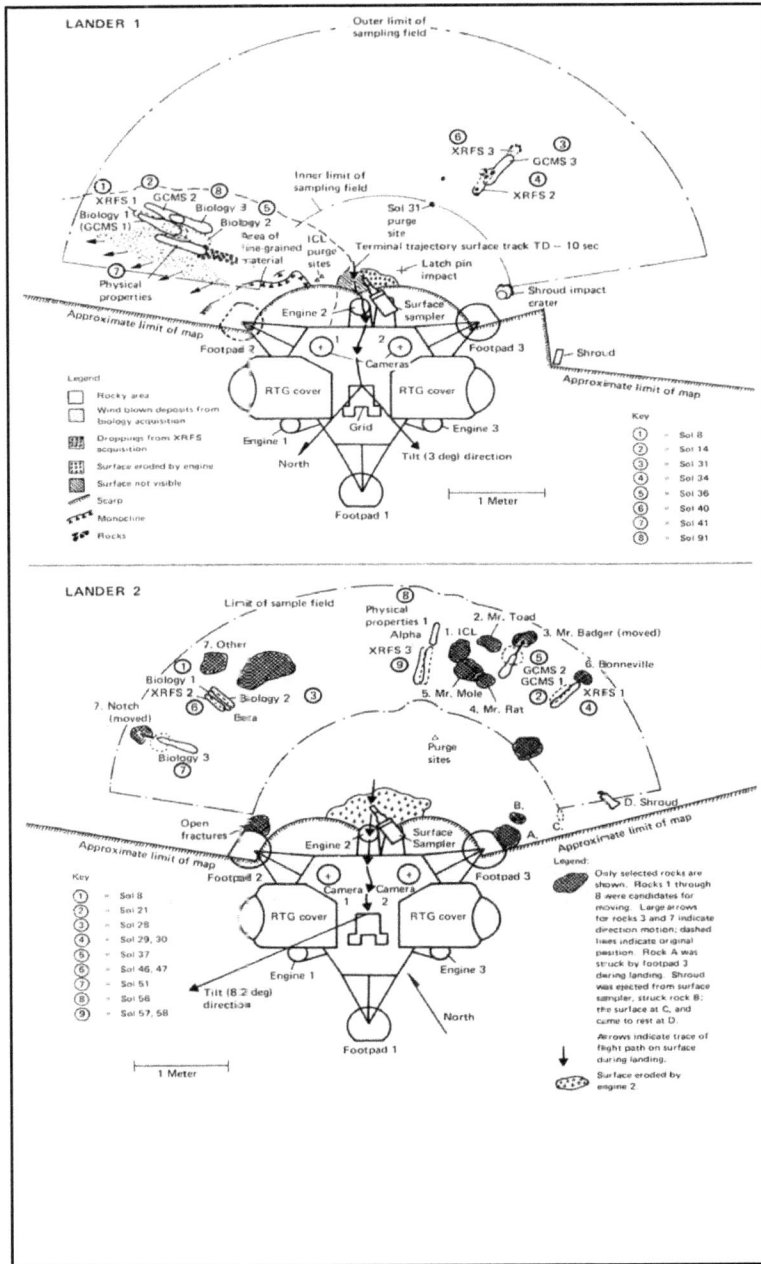

Figure 23. Schematic drawings of Lander 1 and 2 sampling areas summarizing surface sampler activities during the primary mission. The numbered circles indicate, in sequence, the sample acquisitions. The key indicates the sol (Martian day) on which each acquisition occurred.

The Extended Mission

Mars reappeared from behind the Sun, as viewed from Earth in mid-December 1976. This signaled the start of the so-called extended mission. Communication with all four spacecraft was reestablished. This phase of the mission was planned to continue through May of 1978, perhaps even longer.

The extended mission illustrates the general proposition that once a demanding problem has been solved – albeit with great difficulty – the second time is routine. Examples abound in the annals of exploration, particularly mountaineering. So it is with the extended mission. Complicated sequences of trenching and sample acquisition are carried out without fanfare. Many more pictures have been taken than during the primary mission. Before the Viking mission is completed we will have monitored the environment for almost a complete Martian year, or one revolution about the Sun, a total of 687 Earth days. The scientific return is just now being cataloged. It will be years before the analysis is complete. In that sense, the Viking mission is still in its infancy.

Many of the persons who played an instrumental role in the primary mission have now gone. The new team is no less competent, but the excitement inevitably has been replaced by a workaday approach. I return to JPL frequently to examine the most recent collection of pictures, but I feel out of place, like a college graduate who returns to his campus and finds himself among a new generation, perhaps only a few years younger but still a world apart.

The Future

I was encouraged by a valued editorial advisor to close with a look toward the future. In fact, the events of the future are blurred. On a personal basis I realize that many of us who worked on Viking will never again be associated with a comparable journey of space exploration, not because we are in any way disenchanted but because it is time to move on to other things. That separation is bittersweet. The memories keep popping up at unexpected moments, and present ventures seem drab by comparison.

As to further unmanned exploration of Mars, preliminary plans exist for a rover mission in the late 1980s. A tractor-drive vehicle, slightly larger than Viking, could roam up to several hundred kilometers, sampling geological and biological environments inaccessible to Viking. This might be followed by an unmanned sample-return mission in the early 1990s, providing the first opportunity to analyze Martian surface materials in great detail. But these missions exist only on paper. Neither has been approved and their large cost necessarily makes them vulnerable.

Even if the immediate future is uncertain, I have no doubts about the distant years. Some day man will roam the surface of Mars. Those wonderful Viking machines will be crated up, returned to Earth, and placed in a museum. Children in generations to come will stand before them and struggle to imagine the way it was on that first journey to Mars.

Viking 1 Lander Pictures

The pictures in this volume were selected and arranged by Kenneth L. Jones who, in addition, was responsible for the special processing of the original imaging data.

Because the Viking Lander cameras have a fixed vertical field of view of either 20° (high resolution mode) or 60° (survey, color, and infrared modes), but an adjustable azimuthal dimension, many of the pictures are relatively long panoramas. In some instances we have been obliged to split these panoramas so that two parts appear on facing pages. Where this has been necessary we have retained a small amount of overlapping image so that the reader can mentally piece the two parts together.

Individual captions are not intended to be exhaustive descriptions of the pictures. Many objects appear in more than one picture, but are mentioned only where they are shown to best advantage. All particulars relating to camera settings, time of day, picture location, and the like are contained in the table which appears on page 154. The position and scale of pictures can be visualized by reference to the skyline drawings that accompany the table. By using this supporting information, readers can answer for themselves certain technical questions raised by the pictures but not discussed in the captions.

The First Picture

Figure 24

Figure 24 comprises the first picture taken on the surface of Mars. The camera began scanning the scene 25 seconds after touchdown and continued to scan for five minutes. The picture was assembled from left to right during the 20 minutes it took to transmit the data from the Orbiter relay station to Earth. The first segment to be displayed was a narrow strip at the far left. About all that could be determined was the presence of bright and dark areas in the scene, but even that was cause for elation. Ironically, some viewers were more impressed by the picture of the footpad than by the view of the Martian surface, marveling at the fidelity with which the rivets were displayed. The lower edge of the picture is at a slant range of about 1.5 meters, the upper edge about 2 meters. The larger rocks are about 10 cm across. Refer to Figure 19 for location of the picture relative to the spacecraft. The vertical streaking in the left quarter of the picture stimulated a variety of explanations. Those of us familiar with camera operation doubted that it represented a camera malfunction. Instead, something was causing the light levels to vary during the first 1½ minutes following touchdown. It was suggested that clouds were passing in front of the Sun, or, more improbably, that the deployed parachute was casting a shadow as it drifted between the Sun and the Lander. It seemed most likely that dust, kicked up at the time of the landing, was briefly entrained in the lower atmosphere between the camera and the surface. This argument was strengthened by the observation of a sizable accumulation of dirt in the upper concave part of the footpad. Demonstration of the transient nature of the effect is provided by a later picture of the same area (Figure 25) taken with the same approximate lighting. Note that the streaks have disappeared.

Figure 25

Surveying the Landscape

Figure 26 shows the first segment of a 300° low resolution panorama, the second picture acquired from the surface of Mars. For the location of the picture refer to Figure 19.

This image was initiated 6 min after touchdown. The camera continued to scan the landscape for 9 min, at which time it automatically returned to the stowed position. As we viewed the picture, building from left to right, our attention was first drawn to successful deployment of the meteorology boom seen at the upper left. Drifts of sediment are visible in the mid-field. A succession of ridges and depressions lead to the distant horizon.

The bright sky was a surprise. Most scientists had expected that, because the Martian atmosphere is very thin, there would be little scattering of light, and that the sky would be deep blue to black. What we failed to anticipate was a constant pall of fine grained soil particles suspended in the lower atmosphere. These particles scatter incoming light, a large amount of which is directed toward the viewer. The general effect has been described, not very romantically, as similar to a smoggy day in Los Angeles.

The brightness bands in the sky are a camera effect. In fact, the sky continuously darkens at higher elevations, but the camera separates this continuum into a series of sampling intervals (gray levels).

Bright patches and horizontal filaments appear in the upper left of Figure 26, and are more prominent in Figures 27 and 28. They superficially look like clouds. However, they are actually the result of spurious reflections of light from the outer camera housing onto the mirror. We confirmed this theory by producing the same effect in tests on Earth.

Figure 26

Figure 27

As remarkable as the first panorama was, the most dramatic part of the scene was contained in the 60° not covered. Only when we took an equivalent panorama with the other camera three days after touchdown (Figure 27), did we realize what we had missed on the first day – an impressive array large, distinctive boulder.

Figure 28 is a picture taken with the "blue" diode, recording brightness at the shorter wavelengths in the visible range. The meteorology boom appears in the foreground. The bright lower atmosphere is dramatically accentuated.

Figure 29

Figure 28

Desert Drifts

It is entertaining to compare these three views (Figures 29, 31, and 32) of the same region. The first was taken near noon, the second in early morning, and the third in late afternoon. Because of changes in lighting, each picture reveals different features within the scene.

Figure 29 (continued)

Figure 30

Figure 30 should be compared with Figure 29. The two were taken some nine months apart. The lighting is approximately similar. However, the details are not the same. Refer to the large boulder, nicknamed "Big Joe." Just below it, at the right, there is a small region of slumped soil visible in Figure 30 but not in Figure 29. For a geologist, trained to study the changing shape of planetary surfaces, this event is unusually exciting. It represents the first

Figure 32

evidence of contemporary geomorphological change that we have witnessed on any planet other than Earth, or on the Moon. As insignificant as this one small slump is, if the event were repeated every few years, or even every few hundred years, the cumulative effect over geological periods, measured in millions of years, would be pervasive.

Figure 31 was the first of the three panoramas to be obtained. The rising Sun backlights the entire scene, sharply delineating drifts of sediment and shadowing a prominent boulder about 2 m across and 9 m from the spacecraft. This is probably the most publicized picture taken during the entire Viking mission. Within a day after it was released it appeared on the front page of virtually every major paper in the United States, and many other papers around the world. One editor, determined to reproduce the long panorama at optimum size, rotated the picture 90° and printed it sideways. The ultimate compliment came from a friend of mine who could look back on a distinguished career as a photographer for Life during the heyday of that magazine. After the picture was first described at a special news conference, he came up and remarked, "That's a good picture." "Of course," I responded, thinking primarily of its technical qualities. "No," he returned. "You don't understand. It's really a good picture."

Figure 31

Figure 32 (continued)

The origin of Big Joe has not been established. The rock is coarsely granular and is banded in one, possibly two, directions. It may be a breccia fragment, thrown out by a nearby meteoroid impact. Supporting this interpretation is the fact that Big Joe actually is two closely matched boulders side by side. The cleft between the boulders is visible as a notch in the shadow (Figure 32). The two-boulder occurrence appears more consistent with splitting on impact than with erosion in place. Also, some analysts have remarked on the apparent ring-like arrangement of small blocks to the right of Big Joe. However, detailed mapping of the blocks fails to document any circular geometry.

Note that Big Joe is capped with fine-grained material. This may be a remnant of a much thicker cover of dust that formerly covered the entire region to a depth of several meters.

Sculpted Layers

Figures 33 and 34 are two views of sediment drifts showing internal stratification. At first glance these appear to be sets of small scale surface ripples. Closer examination reveals that the linear markings result from erosion of a sequence of sedimentary layers. The same sort of internal stratification can be seen on a seashore beach by trenching through the sand. On a much grander scale, internal stratification of desert dunes is revealed by erosion of Mesozoic rocks in the American Southwest, notably in Bryce and Zion National Parks. In Figure 34 the drift at the right of the image displays several prominent ridges. Darker, more resistant layers are revealed on the eroded drift faces, covered by a thin layer of brighter material.

The presence of sedimentary layers within the Martian drifts indicates a number of depositional events, each related to a change in the wind conditions. The exposure of the layers indicates that the most recent event is general erosion. If these were actively growing dunes the internal stratification would not be revealed.

Four views of isolated drifts of sediment about 20 m to the southeast of Lander 1 are shown in Figures 35 to 38. One is tempted to call these features sand dunes but the label carries certain implications that cannot be established. First, the constituent particles probably are not sand, but are finer grained silt and clay. Second, it does not appear that these are actively growing dunes of the type one commonly associates with deserts. Instead, they appear to be stabilized erosional remnants of a formerly more extensive sediment cover.

Figure 36 was taken with diode BB 3 (high resolution, focused at 4.5 m). Better definition of distant objects is rendered in Figures 35, 37, and 38, taken with BB 4 (high resolution, focused at 13.3 m). Figures 35 and 37 were taken at approximately the same early morning hour. Figures 36 and 38 were taken near local noon; Sun elevation is about 50°.

The same dark drift that appears in the upper right of Figures 35 and 36 is situated in the middle of Figures 37 and 38. A much brighter drift is observable in the upper left of Figures 35 and 36. The difference in brightness (albedo) suggests a compositional difference. This is substantiated by spectral curves generated from data in the three visible color and three infrared channels. The shape of the spectral reflection curves indicates that the dark dune is enriched in ferrous materials, perhaps the mineral magnetite. Pictures acquired much later in the mission, more than a year after touchdown, show dark patches developing on some bright drifts. This suggests that the bright drifts are, in fact, dark drifts with a thin veneer of bright fine grained dust. Generally analogous situations occur on Earth where grains of heavier minerals such as magnetite are preferentially enriched in certain regions where lighter particles of quartz and feldspar have been swept away by wind.

Near the bottom center in Figure 36 there is a cylinder shaped rock. In early news briefings this was nicknamed the "Midas Muffler" rock. However, the same rock as viewed in Figure 35 is not so striking – less of a muffler, more of a garden variety boulder.

The larger boulders on the horizon are about 90 m distant and about 1 to 2 m in diameter. They appear to be eroded into unusual scalloped shapes, although this may be an illusion rendered by low resolution and high contrast lighting. Finer sediment is banked up within the boulder cluster. The same association of boulders and accumulation of sediment occurs elsewhere in the scene. Knobby bedrock is visible in the foreground, and is described in more detail in the text with Figures 52 through 55.

Figure 33

Figure 34

The Martian landscape is characterized by blocks that litter the surface. The interest stimulated by these blocks depends on your point of view. Viewers lacking in imagination see only a rock pile, a sort of cosmic junk heap. At the other extreme, viewers with a surplus of imagination see all sorts of artifacts, even remnants of former civilizations. Somewhere in between, planetary scientists are impressed by block shape, texture, and color – all clues to rock origin and erosional modification.

Figure 35

Figure 36

Figure 37

Figure 38

Blocks

Figure 39

On the Moon, most blocks are formed by meteoroid impact and accompanying excavation of bedrock. Since impact craters are common features on Mars, this same process of block formation must be represented. In fact, Viking Orbiter images of the Lander 1 site indicate that the rims of several impact craters are visible in Lander images, and may have been responsible for much of the apparent fracturing of bedrock in this region. However, at least some of the blocks appear to have modified further by in situ weathering, and moved downslope by processes other than impacts. On a more speculative level, it is possible that some of the blocks were deposited by giant floods that scoured the surface millions of years ago. Braided valleys, similar to those formed by terrestrial flood waters, are revealed in Orbiter pictures as close as 60 km to the west of the landing site.

Whatever their origin, the blocks make fascinating pictorial designs, especially when deeply shadowed by an early morning or late afternoon Sun. Figure 41, taken during the extended mission when a global dust storm darkened the atmosphere, gives quite a different impression than that given by Figures 39 and 40.

Figure 42 and morning frontlighting in Figure 43. A low gain antenna used to receive commands sent from Earth to Mars is visible in the lower right. This receiver failed shortly after landing, one of the very few component failures on the spacecraft. Subsequently, all commands were received through the dish shaped high gain antenna mounted at the rear of the spacecraft.

Figure 40

Figure 41

Figure 42

Figure 43

Figure 44

Figure 45

More blocks! Figures 42 and 43 show the same scene under different lighting conditions: afternoon backlighting in Figures 44 and 45 show the same scene with low and high Sun illumination. The large block in the upper center, like many other blocks throughout the scene, appears to be coarsely granular. This type of block may be derived from igneous rock that crystallized at shallow depth within the crust where cooling rates were slow enough to allow growth of large crystals. The pitted appearance is a result of preferential erosion around these grains by wind blown dust.

Figure 46

Figure 47

Many of the blocks seen here, and in previous Figures, have fillets of fine grained sediment preferentially arranged in one direction. These "tails" probably were formed as the prevailing wind transported fine particles. Eroded from exposed regions, the particles were preferentially deposited in the protected lee of boulders. Alternatively, some of the tails may be protected remnants of a dust cover that originally covered the blocks.

Figures 46 and 47 graphically illustrate how the appearance of the scene changes with differing illumination. Erosional depressions around blocks are clearly shown in Figure 46, although little can be deciphered regarding the configuration or texture of the blocks themselves. Figure 47 shows structure on the block in the foreground that has earned it the nickname "Sponge Rock." This distinctive texture must be telling us something. But what? Without actually picking up the rock, examining it with a hand lens or under a microscope – common geological techniques on Earth – it is difficult to make a unique determination. For example, it has been suggested that the darker spots are dark clasts in a lighter ground mass. The texture, then, would be that of a breccia, a fragmental rock formed by successive breaking apart and lithification during repeated meteoroid impact events. Alternately, it has been proposed that the appearance of the rock is due primarily to an unusually porous texture such as forms in some extrusive volcanic rocks.

The erosional collar is a common feature in terrestrial deserts where similar depressions occur on the upwind side of boulders. The wind, driven against the boulder, develops strong turbulence and increased erosive power. Consequently, fine grained material is scoured in front of the rock.

The elliptical pits in the sediment were formed at the time of landing by throw-out of pebbles or soil clods.

Figures 48 and 49 are two views of the same area, looking back across the spacecraft. The prominent structure on the right is the supporting mast for the S band antenna. Nicely framed by obscuring parts of the spacecraft is a coarsely pitted block, partly buried by sediment. The sediment may represent a large wind tail similar to smaller deposits of sediment that occur in the lee of many small blocks photographed in front of the spacecraft.

The upper part of the strut assembly for the leg at the rear of the spacecraft appears at the left of Figures 48 and 49. Note the striped rod which measures the amount of compression in the landing shock absorber, roughly analogous to an extended, indexed rod in an air pressure gauge for tires.

Figure 48

The diversity of rock types of the Viking 1 landing site has been a matter of controversy. Some geologists argue that only one or two fundamental rock types are represented, and that the diversity of block shape and size is attributable to erosion. Other geologists claim to see ten or more fundamentally different rock types, indicating igneous rocks of varying chemical composition crystallizing under varying conditions. Figure 50 shows a variation in block type that is beyond dispute. In the mid-field there is a group of dark boulders that contrast with the brighter boulders elsewhere in the scene. Figure 51 is an enlargement of part of the scene showing the characteristically pyramidal shape of the dark blocks. Multifaceted blocks of this type commonly occur in terrestrial deserts. They are termed ventifacts, literally "made by the wind." Prevailing winds sandblast one face. Then either the block or wind shifts to create another face. The planar perfection of ventifacts depends on grain size. Fine grained rocks such as basalt produce unusually well formed ventifacts. Coarse grained rocks such as granite tend to yield irregularly pockmarked faces. It is probable, then, that the dark, faceted Martian blocks are wind shaped boulders of basalt.

Figure 49

Figure 50

Figure 51

Outcrops

Bedrock exposures are among the most provocative features at the Viking 1 landing site. Bedrock, of course, refers to a body of solid rock that underlies a layer of soil or unconsolidated sediment. On the Moon, bedrock is almost everywhere covered by a thick layer of impact debris. Only at the Apollo 15 site were astronauts Dave Scott and Jim Irwin able to chip off samples of bedrock from layers exposed in the wall of the Hadley Rille.

Figures 52 and 53 show a knobby exposure of bedrock broken by vertical fractures. Some of the protuberances are so deeply eroded that they almost form detached boulders. Indeed, we believe that many of the boulders at this site form by this mechanism of in situ weathering.

The area of bedrock in Figures 54 and 55 is more clearly delineated. Horizontal rock surfaces have been swept clear of sediment. The fact that this rock weathers in a different fashion than the rock displayed in Figures 52 and 53 suggests a compositional difference between the two exposures of bedrock. Both rock types are believed to be representative of volcanic lavas that inundated the entire region early in Martian history.

Figure 52

Figure 53

Figure 54

Figure 55

Finding bedrock at this landing site was a surprise. Among other things, it indicates that Martian geologic history is more complex than lunar history. If a thick layer of ejected deposits once covered this region, it has been subsequently removed by wind, water, or maybe even ice in the form of glaciers.

At the right edge of Figure 53 is a block with a marking that, generously interpreted, is the letter "B." An enlargement of the same block is shown in an inset.

This block was first photographed approximately five days after landing. The marking elicited no comment from scientists, but one of the media representatives excitedly reported graffiti on Mars. A press conference was hastily arranged, at which time geologists explained that there are many surface markings and erosional stains on rocks that superficially resemble organized symbols No doubt, some viewers with vivid imaginations remain unconvinced.

After the letter "B" was first discovered, scientists amused themselves by searching for other letters. Commonly, the letters they discerned happened to be the first letter of their first or last names. Psychologists take note!

Landing Scars

Although the Viking 1 landing was very gentle, examination of subsequent pictures revealed more and more signs of surface alteration at the time of landing. Many of the changes were due to the exhaust of the three small retrorockets mounted one on each side of the triangular Lander body. Figure 56 shows, on the left, an area beneath the retrorocket mounted between the two cameras. A region of flat, indurated, cracked material has apparently been swept clear of overlying sediment. Similar dried and mud cracked layers occur in arid regions of Earth where they are sometimes referred to as hardpan or caliche. The layers are indurated by the precipitation of cementing salts as soil water, rising to the surface by capillary action, evaporates. We termed the Martian cemented soil layers duricrust, a general term with a minimum of genetic implications. Although the interpretation of this particular exposure was greeted skeptically by many scientists, subsequent observations of similar cemented layers throughout both landing areas strengthened the initial arguments.

Figure 56

Figure 57

Figure 56 (continued)

A small cylindrical pin rests on the soil, about halfway across the bottom of Figure 56. This is the infamous locking pin that should have been ejected from the surface sampler during its first maneuver. Because the arm was not extended far enough the pin failed to drop to the ground, and the sampler mechanism jammed. Subsequently, when the arm was commanded to extend further, the pin finally dropped out. The proof is in this picture.

Figure 58

Figure 57 shows a common feature wherever fine grained material is present. Small pebbles or pellets of soil thrown out by the force of the rocket exhaust have fallen back into fluffy sediment, creating elongate pits. This sort of feature occurs at many different scales on planetary surfaces. Some large lunar craters such as the Imbrium impact basin have secondary craters that extend over an entire lunar hemisphere.

Figures 58 and 59 are panoramas showing the same region to the left of the spacecraft, looking toward the front. At the far left there is a strut leading down to footpad 2. The footpad, itself, is buried by sediment, and is therefore not visible.

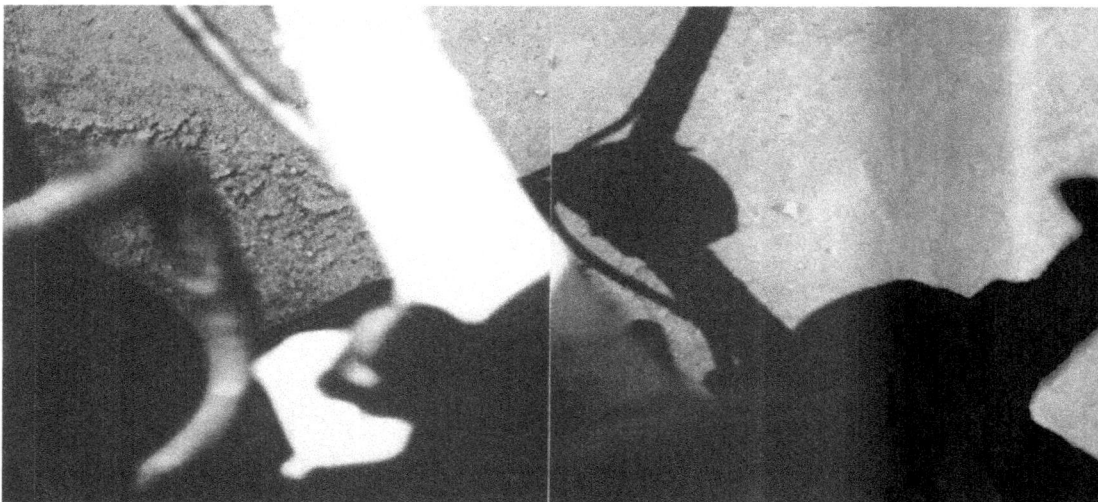

Figure 59

For a landing site purposely chosen as bland and homogeneous, these pictures document surprising diversity. One footpad, that photographed in the very first picture, is perched on a hard, unyielding surface. Another footpad, less than 3 m distant, is emplaced in a drift of incohesive, fine grained sediment. At the time of landing this sediment broke away and flowed down a gentle slope to cover the footpad.

Elliptical pits, formed by rocket exhaust throw-out of soil pellets, are visible in the right part of the panorama. The prominent structure at the far right is a magnet cleaning brush. Permanent magnets were mounted on a back hoe attachment to the sampling shovel. By dragging the back hoe through the soil the concentration of magnetic particles could be determined. The initial plan was to clean the magnets after each sampling operation by moving the back hoe between the opposing stiff wire brushes on the end of the struts. During tests conducted on Earth it was discovered that the surface sampler occasionally became jammed between the wire brushes. Accordingly it was decided that the cleaning brushes would not be used. Eventually, well into the extended mission, the edict was rescinded and several cleaning operations were successfully accomplished.

Figures 60 and 62 are two pictures of the same area taken under different lighting conditions, although Figure 60 shows a larger area than does Figure 62. Figure 61 is an enlargement of the area in the upper right corner of Figure 60. Pits formed by throw-out of soil pellets during landing rocket maneuvers are particularly well displayed in the right side of Figure 62. At the far right of Figure 61, just above the landing leg strut, a beer can shaped object is visible lying on the surface. This is a shroud that protected the surface sampler head and was ejected after landing, bouncing, and rolling to its present position. The apparent coarse texture of the sediment in the upper left of Figure 61 may be an artificial effect, caused by "aliasing." Aliasing results when periodic small scale brightness variations in the scene cause systematic variations in larger picture elements.

Figure 60

Figure 61

Figure 63 shows a rectilinear set of ridges and cracks in fine grained material. Apparently the sediment is cohesive enough that it breaks up along discrete planes. Its properties have been compared to that of wet sand, although the moisture content of the Martian sediment is exceedingly low.

Figure 62

Figure 63

Behind the Lander

Figure 64

During the planning stages for the Viking mission, we generally discounted the field of view looking back across the spacecraft. Only small fragments of the Martian surface would be visible, and, for that reason, the pictures offered little promise.

In fact, some of the most provocative pictures acquired during the mission show the spacecraft in the foreground (Figures 64 and 65). The effect is essentially that established by any traveler when, in taking a photograph of some landscape, he has his companion stand in the foreground. In so doing he seeks to emphasize his personal experience of first hand viewing. So it is with our journey to Mars. By itself, a picture of the Martian landscape has no frame of reference. But the spacecraft in the foreground imparts an immediacy, a reality. We recognize the familiar components, assembled on Earth and now situated on Mars. From this point of recognition it takes only a small jump of imagination to place ourselves on Mars.

Figure 65

Figure 66

Figure 67

These panoramas show the same field of view, looking northwest. Figure 66 was taken in the morning, Figure 67 in mid afternoon. The wind cover over the radioisotope thermoelectric generator (RTG) is visible in the foreground. Scalloped drifts or sediment and clusters of boulders accent the landscape. As in other parts of the scene the larger boulders are localized on small hillocks and appear to be either loci for the accumulation of sediment, or stabilizing

Figure 68

Figure 66 (continued)

elements for residual sediment. The small parallel ridges in the left half of Figure 66 superficially resemble wind generated ripple marks, but are probably caused by exposed layering in the drift material.

This striking panorama shown in two parts (Figures 68 and 69) covers approximately 100° in azimuth. The scene is backlit. The late afternoon Sun is about 18° above the horizon near the middle of the panorama. Some objects on the horizon, as in Figure 69, sparkle with reflected light. The top of the low gain antenna is shown in Figure 68. The prominent structure visible in Figures 69 and 70 is the wind cover for the RTG.

The symmetrical ridge on the horizon at the right of Figure 68 is a crater rim. It is further described in the text accompanying Figures 77 through 79. The origin of the prominent trough in the foreground is undetermined. The higher mound of material in the right half of Figure 68 is about 30 to 60 m from the spacecraft. In fact, it is a complex series of ridges occurring at increasing distances from the spacecraft.

A part of the same scene, photographed with early morning light, is shown in Figure 70.

Figure 68 (continued)

Figure 69

Figure 70

These Figures dramatically demonstrate the effect of different lighting in revealing detail within the scene. Figure 71 was taken during midday, Figure 72 in late afternoon, and Figure 73 in early morning. Some of the boulders closest to the spacecraft may have been sculpted by wind driven sand. Particularly well shown in Figure 73 are ripple like alternations of bright and dark sediment, a probable consequence of wind sorting and layering during transport.

Numerous pits are visible, especially in Figure 73. These are caused by material kicked from beneath the Lander during touchdown.

A part of the UHF antenna that relays Lander data to the Orbiter is shown at the far left. The flat surface at the bottom of the picture is the top of a protective box covering the seismometer. The seismometer on Lander 1 failed to operate, the only significant instrumental failure on either spacecraft. Although the seismometer on Lander 2 worked as anticipated, its ability to detect Marsquakes was greatly compromised because it was mounted on the upper deck of the spacecraft. Most of the recorded vibrations were induced by wind or by operation of spacecraft components.

Figure 71

Three views (Figures 74 to 76) looking over the Lander through a forest of tubes, most of which served as umbilicals cords linking the Lander to the Orbiter during the 10 month journey from Earth to Mars. The unusual graphic effects are reminiscent of a science fiction illustrator's vision of what a spacecraft on an alien planet should look like.

Particularly well displayed in Figures 74 and 76 is a view of the same drift complex shown in Figures 66 and 67, although this view is taken with the other camera.

These are views looking toward the southwest horizon at different times of day: early morning (Figure 77), early afternoon (Figure 78), and late afternoon (Figure 79). Different lighting accentuates different detail in both nearfield and farfield. Of special significance is the symmetrical skyline ridge revealed in Figures 77, 78, and 79. Almost certainly this is the upraised rim of an impact crater, about 500 m in diameter and 2.5 km distant. Part of the far side of the central cavity can be discerned toward the left, particularly in Figure 79.

This same crater has been identified in Orbiter photographs and has been used to locate the Lander position. Refer to Figure 204. Also visible beyond the crater rim is a ridge 8 to 10 km distant.

Figure 80 is the first picture taken by Viking 2 Lander. The preprogrammed picture sequence for the first several days was nearly identical to that stored in the Lander 1 computers (Figure 19); to that extent, certain aspects of the Viking 1 and 2 pictures are similar. Note, for example, the vertical streaking at the left side of Figure 80. In general, this is very similar to the streaking in the first Viking Lander 1 picture. It occurs in none of the more than 3000 other pictures taken throughout the mission, including a later view of the same area (Figure 81). This evidence conclusively demonstrates that the streaking is the result of a singular event – raising of a dust cloud at the time of touchdown. The detailed configuration of streaking at the two landing sites is different, indicating slightly different dissipation of the dust. At the Lander 1 site the atmosphere was clear after 1 1/2 min. At the Lander 2 site the dust settled more rapidly; clearing was complete within one min.

Figure 72

Figure 74

Figure 73

Figure 75

Figure 78

Figure 79

Figure 79 (continued)

Figure 78 (continued)

Figure 76

Figure 77

Viking 2 Lander Pictures

Another First

We had expected (hoped?) that the second landing site would be relatively free of blocks. This first picture made our predictions look a little shaky. The sponge-like rock in the right half of Figure 80 provides a dramatic example of vesicular texture. The subspherical cavities probably formed around small gas bubbles in cooling volcanic lava.

Tilt

Figure 82 is the second picture taken by Lander 2, a low resolution panorama that was completed about 20 min after touchdown. The most striking feature is the sloping horizon. Recall that the picture is assembled from left to right. As we looked at the first few lines all we could tell was that the horizon dipped to the right. Although it was possible that this was an actual feature within the scene, we suspected otherwise. As the image continued and we watched the horizon regularly descend and then climb back up our suspicions were confirmed. The horizon was, in fact, very nearly level but the spacecraft was tilted approximately 8° toward the northwest. The effective result is that in the middle of the panorama, looking toward the southeast, the camera is tilted up and the horizon appears near the bottom of the image. Pointing in the opposite direction northwest, the camera's field of view is tilted downward and the horizon appears near the top of the image.

Since the tilted spacecraft is situated on a nearly level plain, the obvious conclusion was that we were perched on one of the thousands of boulders that littered the plains as far as our camera could see Initially, the health of the spacecraft was in doubt. If a boulder had penetrated the Lander underbelly, then the internal components might have been destroyed by cold during the Martian night. Fortunately this was not the case. When the Orbiter passed overhead 24 hours later, the Lander telemetered data indicating it was alive and well.

Figure 80

Figure 81

Figure 80 (continued)

Figure 82

During landing, sediment collected in the dish shaped S band antenna which was stowed in a concave up position.

When the antenna was rotated 90° the sediment slid to the lower edge. However, interpretation of the image was difficult. Some viewers thought they saw a ragged crack in the antenna.

Note the bright streak in the sky. The fact that it is horizontal, even though the horizon is tilted, indicates that it is not an atmospheric effect but is due to spurious camera reflections, as described in Figure 24. This interpretation is further confirmed by the fact that the "cloud" passes in front of the S band antenna.

A Rocky Plain

Two views of the same rock strewn plain are shown. Figures 83 and 84 were taken in the early morning. The Lander faces approximately northeast. With lighting from the east the S band antenna, mounted on the rear of the spacecraft, casts a shadow on the ground to the left of the Lander. Figure 85 was taken at noon.

The many boulders seem to stand apart from the bland background, as if they were carefully positioned on some giant Hollywood set. This appearance probably is due to erosion by wind – eolian deflation, in geological parlance – that has carried away fine-grained material between larger rocks. Some of the boulders actually stand on pedestals of protected fine grained substrate.

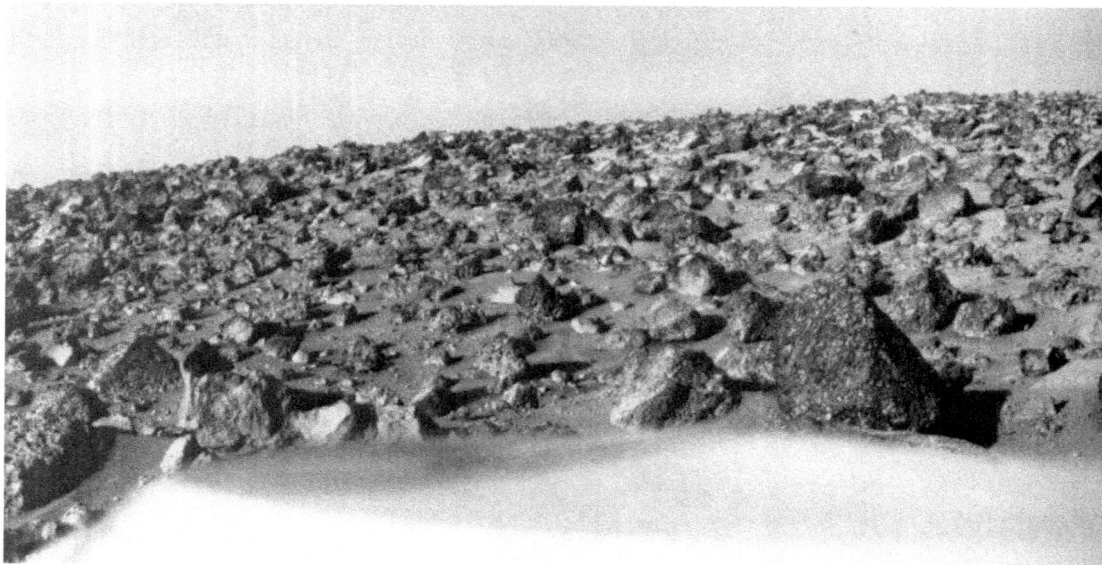

Figure 83

Figure 85

Figure 84

Figure 86

Figure 87

Figure 88

Figure 89

Figures 86 and 87 are two views of the same area in front of the spacecraft looking toward the north to northeast, both taken with camera 2 but at different times of day. Linear accumulations of fine grained sediment have planar surfaces that are sharply delineated by shadow and reflected light. The small pyramidal rock in the lower left is a good example of a ventifact, a rock with multiple facets eroded by the wind.

Figures 88 and 89 are two views of an area in front of the Lander taken with camera 1. The prominent spacecraft structure is the meteorology boom. Note the shallow trough, especially prominent in Figure 89, that can be traced down and to the right. Extending to the right of these pictures it becomes even more prominent, as in Figure 100. Small crescent shaped deposits of fine grained material are visible along the entire length of the trough.

Figures 90 and 91 constitute a single panorama, taken in the early morning. The prominent cylindrical object in the lower right of Figure 91 is a protective shroud that covered the surface sampler during transit to Mars. It was deployed the second day after landing. Its regular outline and brightly reflecting surfaces stand in vivid contrast to the natural scene.

The long early morning shadows consistently extend to the west, but they appear to change direction as the perspective of the viewer changes around this 60° panorama.

Figure 92 shows the area adjacent to the sampler shroud taken with high Sun illumination. The large rock at the top of the image exhibits an unusual scalloped surface, probably the result of wind erosion.

Figures 93, 94, and 95 show views of the surface, with the strut for leg 3 in the foreground, at three different times of day. Note the parallel lineation, probably due to wind scour. Small areas of duricrust appear near the top of the picture. Patches of coarse pebbles throughout the scene may result from removal of fine grained material with attendant concentration of coarser particles.

Early morning (Figure 96) and late afternoon (Figure 97) views of the area in front of the spacecraft looking toward the north are shown. The images overlap on the left with Figure 103 on the right with Figure 105.

Figure 90

Figure 91

Figure 92

Figure 93

Figure 94

Figure 96

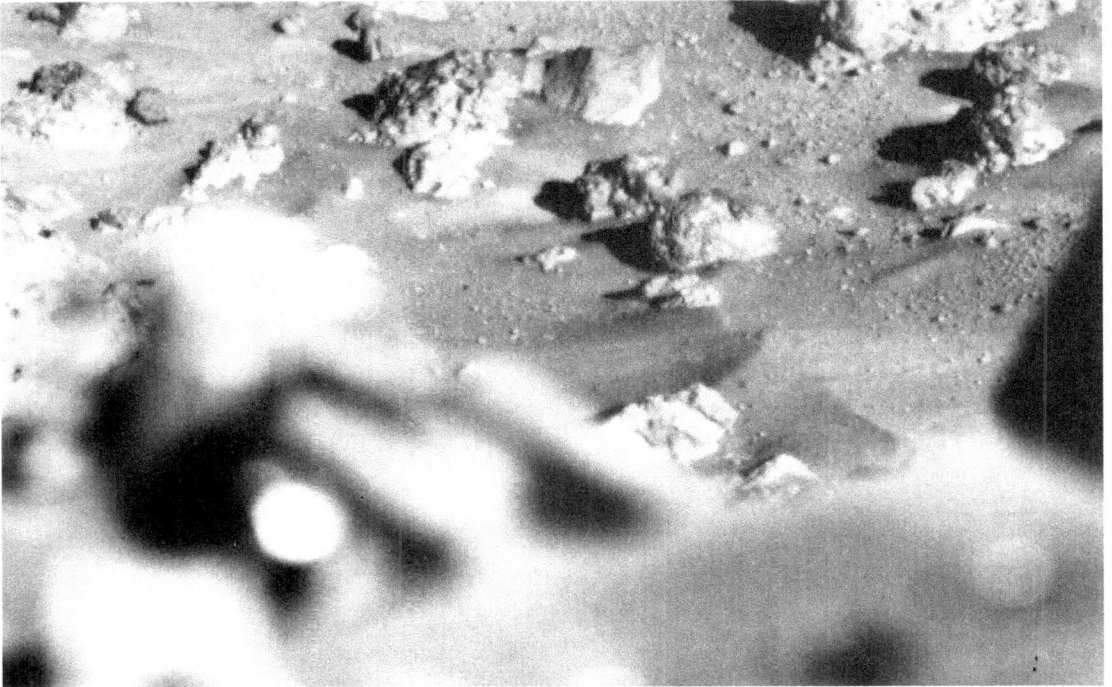

Figure 95

Figure 97

Figure 98

Figure 99

The shadows in Figures 98 and 99 are noteworthy. When the cameras are not in use they are stowed with their transparent windows positioned behind protective posts. The placement of these posts was a matter of great debate. Since they obscure about 20° in azimuth of the field of view it was decided to put them where a minimum of the natural scene would be affected. The optimal position was looking toward the opposite camera. The consequence, of course, is inability to take a picture of one camera with the other. During the months prior to launch, when we suspected we might have trouble with the camera azimuth drive, we wondered whether it might not have been an error to deny ourselves a potentially instructive picture of a malfunctioning camera.

Figures 98 and 99 are the closest we can come to a self portrait. Figure 98, taken with camera 1, shows the tip of the shadow of camera 2, situated between the shadows of the S band antenna and the meteorology boom. Figure 99, taken with camera 2, shows the shadow of camera 1 at the upper right. The shadow of the surface sampler is in the center.

Enigmatic Troughs

Figure 100

Figure 100 is one of the more instructive pictures taken at the Viking 2 site. A linear depression, or trench, can be traced across the middle of the picture. The bottom of the trench is 10 to 15 cm lower than bordering lips. The trench can be traced more than 10 m (Figures 103 to 106), trending generally east west and descending slightly to the east. It is partly filled with sediment finer than on adjacent surfaces. The sediment apparently has been scalloped by the wind. Larger rocks are relatively rare, and the few that are present are partly buried by finer sediment.

The trench seen here is part of a more extensive polygonal network that occurs near the Lander. The origin is uncertain but it is possible that the polygonal soil structure is the result of cyclic freezing and thawing of ground water. Analogous structures form in periglacial regions on Earth. During spring, melt water segregates in the soil and freezes to form an ice wedge. Expansion of permafrost during the summer causes an upward bulge in the vicinity of the ice wedge. During the winter the permafrost contracts and the wedge opens, initiating another annual cycle of wedge growth. A depression forms over the wedge and, in windswept terrains, is filled by wind transported sand.

Figure 101

Figure 102

Viking 2 landed at 48°N, 25° north of the Viking 1 site. This difference in latitude might account for "polar" landforms at one site but not the other. If the trenches do reflect the subsurface formation of ice wedges in the same way as on Earth, then their formation must have occurred at some former time when liquid water was stable. With the present thin atmosphere only the vapor and ice phases are stable. Any liquid water would either freeze or boil away.

Figure 103

Figures 101 and 102 are enlargements of particularly significant regions in Figure 100. The large rock, the left side of which is shown in Figure 101, is approximately 1 m wide. Stratification in the block runs from upper left to lower right. The left end of the boulder is much more pitted than the central part. These pits, which occur in the majority of the boulders, may be the result of vesiculation, or frothing, in a gas charged lava during its consolidation. The layering of pits may represent vertical differentiation in the original volcanic deposit, with the most highly vesiculated lava occurring near the top of the flow.

Figure 102 shows one of the wind sculpted drifts seen at the Lander 2 site. The polygonal shapes indicate that the drifts are probably being eroded by wind activity.

Figures 103 to 106 show the development of the trench as it is traced from left (west) to right (east). The right side of Figure 103 overlaps the left side of Figure 104 which is also displaced downward 10° in elevation. The right side of Figure 104 overlaps with the left side of Figure 106.

Note that many of the features in Figures 104 to 106 are also identifiable in Figure 100. A series of crescent shaped drifts within the trench is particularly well developed. The perspective is different because the scene is photographed with two cameras, situated 0.8 m apart on the spacecraft.

Figure 104

Figure 105

Figure 106

Looking Backward

Figure 107

Figure 108

Figure 109

Figure 110

As was the case for Lander 1, some of the more fascinating Lander 2 pictures are those taken looking back across the spacecraft. Figures 107 and 108 show an early morning and midday view of the same region. The pipes at the right were connected to the Orbiter during the trip to Mars in order to stabilize and monitor the internal Lander environment. The top of leg 1, at the rear of the triangular spacecraft, occurs in the bottom center.

Although most of the blocks on the surface are irregularly broken and pitted, some have smooth, conchoidal fractures.

An example appears at the far left in Figure 108. This morphology is typical for glassy to fine grained volcanic rocks.

Figure 109 is a low resolution view showing a larger area than that contained in the high resolution views of Figures 107 and 108. A large number of spacecraft components are identifiable, including the RTG protective covers, two of three reference test targets for the cameras, and the dish shaped S band antenna.

Brightness contouring in the sky, particularly well developed in Figure 109 but detectable in many other pictures, is caused by increasing sky brightness as the camera moves toward the Sun. The Lander camera transforms this continuous gradation into a series of discrete steps with increasing brightness. The filamentous horizontal bright streaks in the sky are caused by spurious reflections of light from the outer camera housing. When the camera mirror is at certain critical positions, this light interferes with the normal radiance of the scene.

Figures 110 and 111 are panoramas looking toward the southwest at different times of day. The hinge assembly on the S band antenna is prominently shown left of center in both figures. The vertical upper mast appears at the center. The horizon near the left edge of the scene is much closer than at the right, because of the presence of a small rise. One result is that individual blocks are visible on the horizon to the left but not to the right.

Two pairs of views behind the spacecraft are shown. Figures 112 and 113 show a distinctive conchoidally fractured block near the middle of the scene.

Figures 114 and 115 show a round 10 cm magnifying mirror in the foreground. During the mission, the sampler arm was positioned in front of this fixed mirror and the reflected image was photographed. In this way it was possible to get a 5X enlargement of magnetic soil particles adhering to the magnets on the sampler head. In the early morning view (Figure 114) an erosional collar of sediment is discernible around one of the blocks.

Figure 110 (continued)

Figure 111

Figure 112

Figure 111 (continued)

Figure 113

Rock Textures

The deeply textured rocks at the Lander 2 site yield attractive graphic patterns, especially when they are accentuated by low Sun and deep shadows (Figures 116 and 117).

The blocks are so distinctive that one might expect that their origin could be easily determined. However, this turns out not to be the case. The abundant pits are similar to vesicles that occur in terrestrial volcanic rocks derived from gas charged lavas. As the gas rises to the surface, some of the bubbles are essentially frozen in the solidifying lava. Although this is the favored interpretation for the Martian rocks, the pits are larger and more widespread than is typical for terrestrial situations. Alternate interpretations are that the pits mark the former presence of easily eroded clasts or crystals, or that the rock results from partial cementation in an upper soil zone in a manner analogous to that for some tropical regions on Earth.

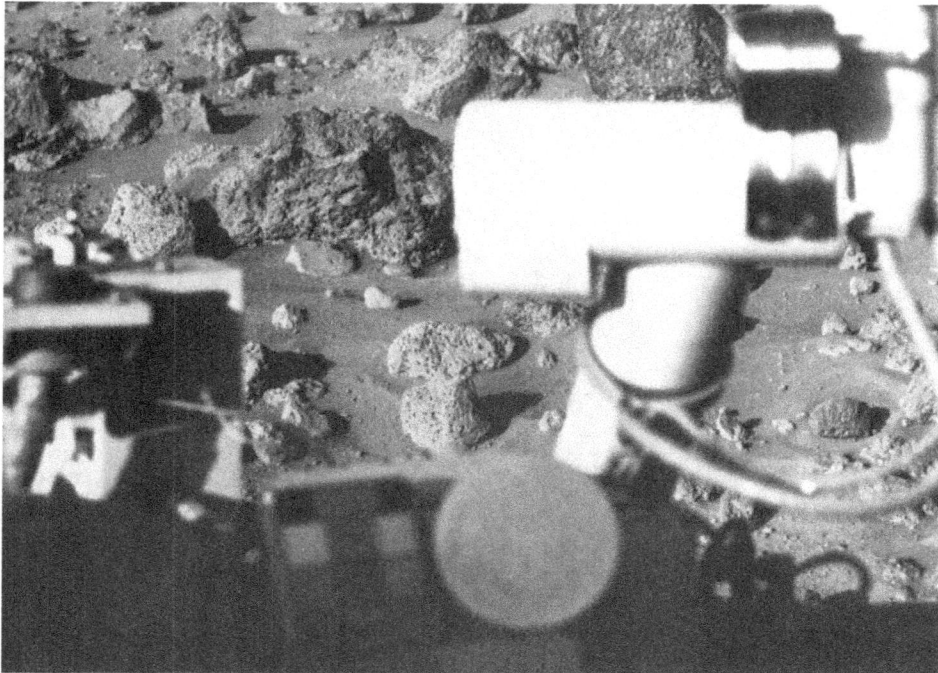

Figure 114

Figure 115

Figure 116

Figure 117

Figure 118

Figure 119

Figure 120

All these arguments proceed by analogy: the features on Mars resemble those on Earth that we know to be formed by particular processes. Unfortunately, it is difficult to quantify the likelihood of one analogous comparison as opposed to another. In reasoning by analogy we are constrained by our own experience. The number of analogies at our disposal is a function of our knowledge of terrestrial land form. If processes on Mars produce unique forms, then dependence upon analogies may blind us to that uniqueness.

Figures 118 to 121 make up two pairs of pictures with different illumination angles. Because of their roughly textured and pitted surfaces, the blocks change dramatically in appearance as the lighting varies from direct to oblique.

Figure 122 is a 100° panorama taken in the late afternoon. The lower part of the meteorology boom appears to the left; the shadow of the upper mast, including sensors, is visible immediately to the right. Figure 122 includes a region from 30° to 50° in camera elevation. Figure 123 is from 40° to 60°, so that a rock that appears in the lower far right of Figure 122 is located in the upper left of Figure 123.

These three figures show the effects of erosion in the vicinity of the descent engine exhaust impact point. A thin veneer of fine grained sediment has been swept away to reveal polygonally fractured "rock." In fact, the polygonal fragments probably represent cemented soil fragments. This inference is strengthened by the observation that chemical analyses of unconsolidated soil closely resemble analyses of the platy chips. Refer to Figure 56 for documentation of a comparable effect at the Lander 1 site.

This indurated soil zone, termed duricrust, has probably formed by upward migration and evaporation of ground water with attendant precipitation of cementing compounds. Although duricrust was speculatively identified at the Viking 1 site, it was not until we received these Viking 2 pictures that we felt confident of the interpretation.

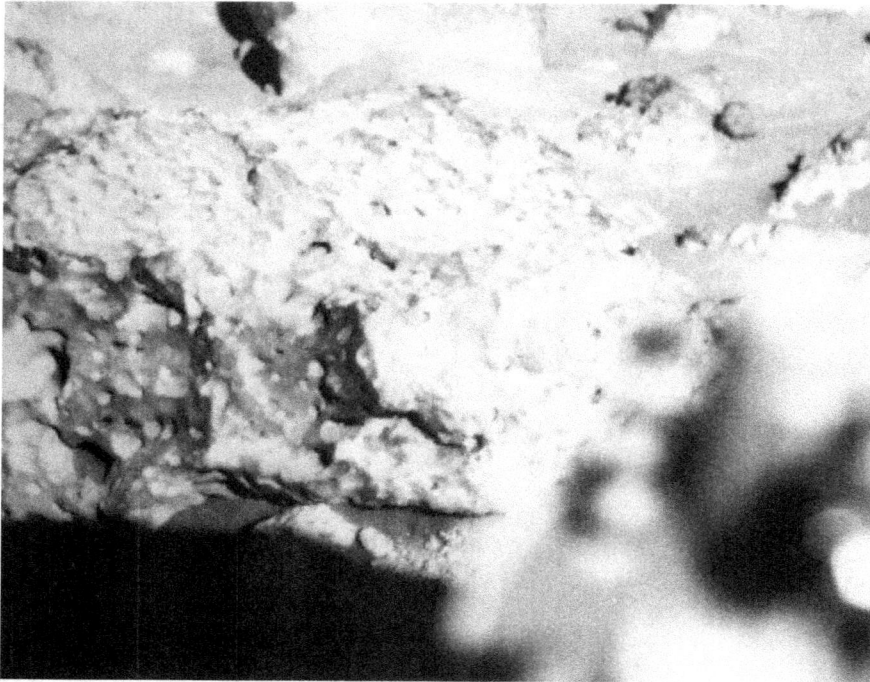

Figure 121

Figure 122

Figure 123

Figure 122 (continued)

The prominent structure in Figures 124 and 125 is the magnet cleaning brush. Mounting brackets for the surface sampler assembly appear at the left in Figure 126.

Figure 124

Figure 125

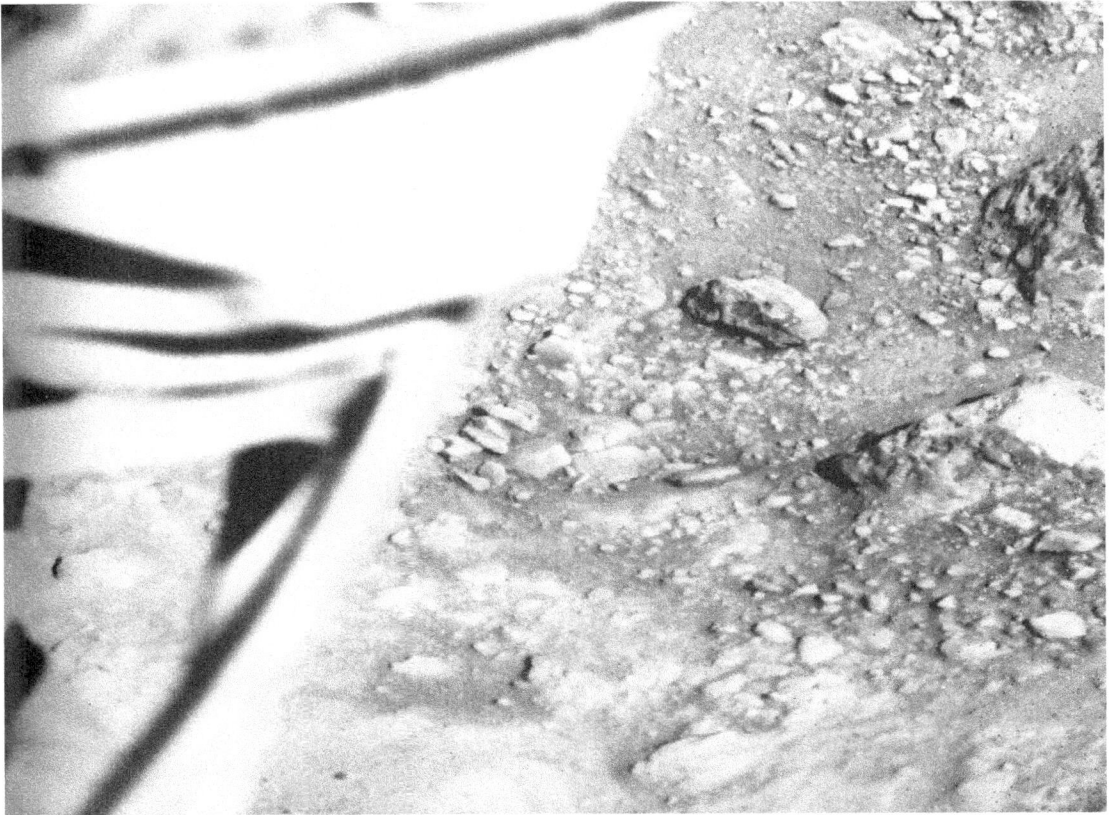

Figure 126

Figure 127

Figure 128

Figure 129

Two views of rocks and sediment at the Viking 2 site. Fillets of sediment that partly bury some blocks are shown well in Figure 127. The grazing early morning light accentuates parallel grooves in the large block at the center of Figure 129. This stratification might have been caused by flow in a viscous volcanic lava.

Approximately vertical illumination in Figure 128 brings out differences in soil texture. The finer grained material appears to be a patchy wind deposit superposed on coarser accumulations of cemented soil fragments. The surface sampler has been unable to acquire any of these coarser fragments which apparently disintegrate when scooped up.

Figure 130

Figure 131

Three views of the same area were taken in late afternoon (Figure 130), early morning (Figure 131), and midday (Figure 132). The appearance of the scene changes dramatically with the change in illumination. Although many of the blocks have the pitted appearance characteristic of the Viking 2 site, some are clearly different in morphology and, perhaps, chemical composition. Look, for example, at the smoothly polished block that appears a little to the right of center in Figure 132. Its singular appearance is more subtly revealed in Figures 130 and 131. In the first case, a smooth face reflects the evening light.

Orthogonal blocks occur throughout the scene. Some wind faceting may have occurred, but most of the blockiness is probably the result of erosion of a volcanic lava that was traversed by several sets of mutually perpendicular joints.

Figure 132

Figure 133

Figure 134

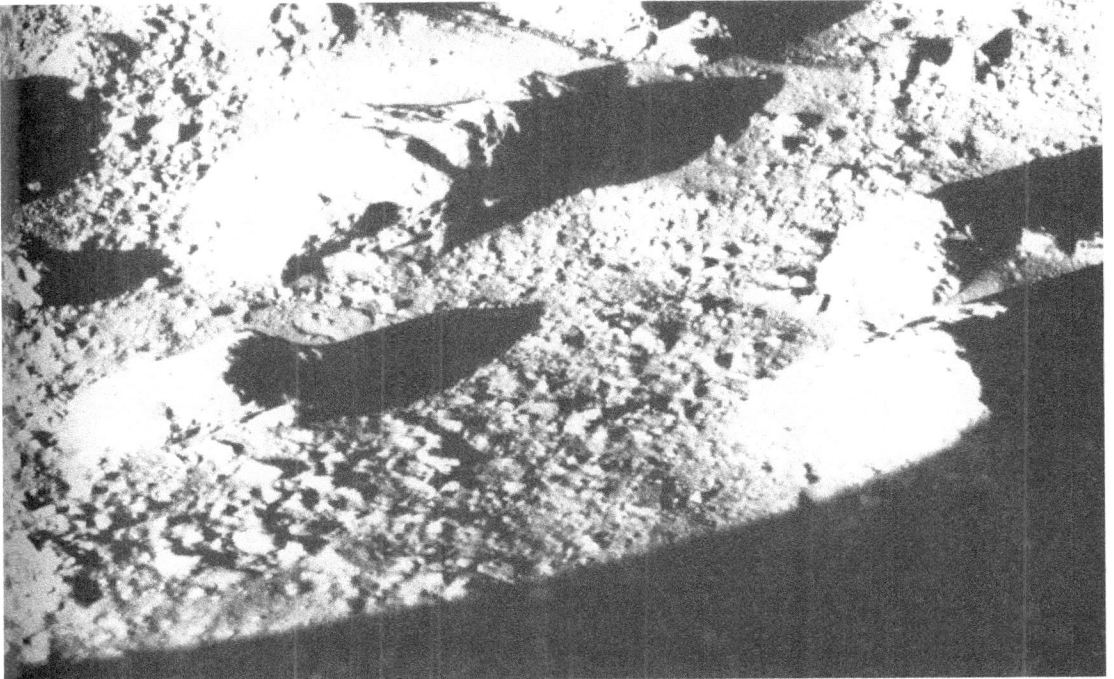

Figure 135

Note two boulders that occur close together a little to the left and above center in Figure 133. They have matching faces which suggest that a single boulder has been split in two and the two parts separated, perhaps by frost heaving. Several of the large vesicular boulders in this scene display parallel stratification. A trench dug by the surface sampler is visible in the lower right.

Oblique lighting in Figure 134 accentuates pits so large they look like small craters. Indeed, if Mars lacked a shielding atmosphere that destroyed small meteoroids, an impact origin for the pits probably would be favored.

Figure 136

Figure 137

The spacecraft shadow in the lower part of Figure 135 creates an attractive graphic design. Note that the right part of Figure 135 is the same field of view as the left center part of Figure 134.

Figure 138

Figure 137 (continued)

The Distant Horizon

This panorama of almost 100°, looking toward the east, is broken into two segments, Figures 136 and 137. In Figure 136 a sequence of increasingly distant ridges can be identified. In the right half of Figure 137 a block studded crest close to the spacecraft obscures the distant horizon. The general slope of the horizon is a consequence of spacecraft tilt, not of any natural gradient in the scene. Indeed, the horizon is almost exactly level, displaying significantly less relief than the distant landscape at the Viking 1 site. By coincidence the farthest horizon appears at the lowest point in the tilted image.

The origin of this gently undulating plain at the Viking 2 site is a matter of conjecture, but a likely possibility is that the Lander is situated on a vast ejecta deposit associated with a 100 km crater, Mie, situated 160 km to the east. Eolian deflation might have stripped much of the fine grained material originally in the ejecta deposit. Unfortunately, no land forms in the Lander camera images are identifiable in Orbiter images.

Figure 139

Figure 140

Figures 138, 139, and 140 show horizon ridges photographed at three different times of day. Predictably, change in illumination results in different reflections from the broad slopes. Only by reference to all the available pictures can a viewer gain a true impression of the complex detail in the distance. Figure 138 is purposely printed dark to bring out the relatively light horizon details.

Special Effects

Calibration Pictures

   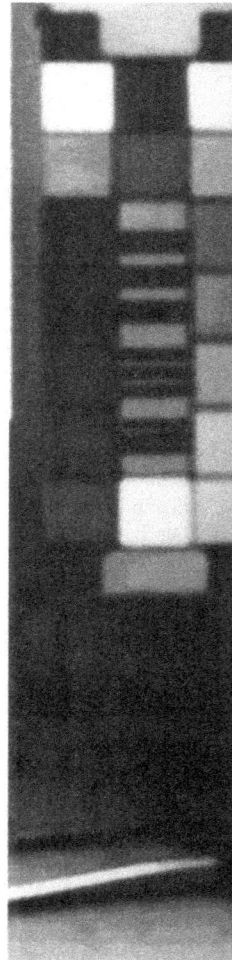

Figure 141   Figure 142   Figure 143   Figure 144

The performance of the Viking cameras can be monitored in several different ways. The sensitivity of all 12 diodes is tested by flooding the photosensor array with light from a bulb directly above the array. The returned data are displayed as a series of bright vertical stripes, each stripe corresponding to the signal from one of the 12 diodes. Although the data are displayed here in conventional camera format (Figure 141), neither the vertical nor horizontal scanning mechanisms are used in this test.

The operation of the scanning mechanisms is tested by imaging two point light sources positioned one above the other in the fixed post against which the camera is stowed. Originally intended to be used prior to touchdown to confirm that the camera was operating properly, scan verification images have been used throughout the mission for completely different purposes. Figure 142 shows a scan verification image taken just before the transparent camera window was exposed to a jet of compressed CO_2; Figure 143 is a second scan verification picture taken just after this event. The similarity of the two pictures suggests that no dust was adhering to the window. If so, the lights in the post would have appeared brighter following the cleaning operation. Quantitative comparisons are made not from images but from computer printouts showing the pixel brightness values.

A more conventional camera test is imaging a target mounted on the spacecraft (Figure 144). Actually there are three identical targets, two of which are visible with either camera. Square patches of red, green, and blue paint assist in establishing color balance. A series of gray chips with varying brightness calibrate sensitivity of camera diodes and associated electronics. Groups of bare with different spacing indicate the camera's spatial resolution, much as we determine the quality of our own vision by reading progressively smaller symbols on an eye chart. A color picture of a test chart is shown in Figure 184.

The properties of the test charts were measured carefully before they were mounted on the Lander, but their usefulness on Mars was limited by the fact that they were shadowed by spacecraft components – particularly the dish shaped S band antenna – through much of the Martian day. During those times it was difficult to use reflectance values for calibration of associated images. In addition, dust accumulation o the charts, both from surface sampler activities and from atmospheric settling, diminished their usefulness.

Something From Nothing

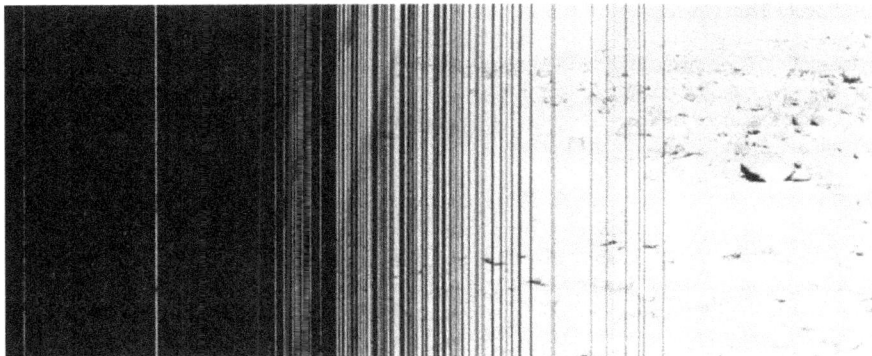

Figure 145

Figure 146

Figure 147

This series of pictures illustrates the way in which so called raw data can be recovered and modified to produce a picture o] optimum quality and usefulness.

Data loss can occur at a number of points as information is transmitted from Lander to Orbiter, from Orbiter to antennae girdling Earth, or from the receiving stations to JPL. Early in the mission all critical data were transmitted redundantly to protect against loss.

In Figure 145 the data loss occurred during the Lander to Orbiter relay. As the Orbiter rose above the horizon, the received signal from the Lander became progressively stronger. The initially weak signal caused a large number of "bit" errors. In many instances the computer was unable to locate the beginning of a line in the telemetry stream. Reprocessing of the data located most of the image data that was present but misaddressed (Figure 146).

There remains a scattering of isolated picture elements with incorrect brightness numbers due to infrequent, but statistically significant, errors in transmission. This is termed "noise," analogous to the background static that interferes with a conventional radio broadcast. Noise is removed by assigning the spurious pixels values intermediate between those of neighboring pixels. Although the apparent quality of the picture is dramatically improved there is only a small gain in actual information.

The contrast of a picture can be increased. In Figure 147 the brightness values of constituent pixels have been "stretched" to encompass a larger range of gray levels. Most of the brightness numbers recorded by the camera were clustered between 8 and 40. To increase contrast the limits of the distribution were reassigned values of 0 and 63, and intervening brightness numbers were stretched linearly. The result is that the bright pixels are brighter while the dark pixels are darker.

Another technique for increasing discriminability involves use of a mathematical construction called a box filter. Individual pixel brightness values are compared to the average brightness within a square array of neighboring pixels. The difference between the two is then increased by an arbitrary factor. Increasing the contrast of fine detail while leaving unchanged or removing larger scale variations has the visual effect of "focusing" the scene.

Searching for Movement

One of the prime objectives of the imaging investigation was to document changes, either rapid or long term, within the scene. Figures 148 through 151 are a series of pictures of footpad 3 at the Viking 1 site. Taken over a period of several days, these pictures record slightly different shadow patterns corresponding to different times in the morning. No other changes are apparent. Similar monitoring sequences were used extensively throughout the mission to photograph regions that seemed the best candidates for change. The most valuable sequence of repeated images documents the formation and decay of a frost cover at the Viking 2 site (Figures 194 to 196).

Figure 148

Figure 149

Images are being acquired that precisely duplicate lighting of early mission pictures. These images indicate that a thin surficial cover of light dust settled out at the Viking 2 Lander site after a major planetwide dust storm.

As previously discussed in the introductory text, the Viking cameras have an unusual capability for detection of motion within the scene. The azimuth motion of the cameras can be inhibited so that the same vertical line is scanned repeatedly. Sequential scans are displayed in the usual way, building up from left to right. The horizontal axis can be considered as time. In Figure 148, for example, the first repeated line scan occurs at the right edge of the conventional image, the last repeated scan at the far right. In all, 43 repeated line scans are shown. This picture was taken at the slower rate of 13.65 seconds per scan. Accordingly, the elapsed time of the monitoring sequence is 587 sec. The same low scan rate was used to monitor slowly changing atmospheric conditions near sunrise or sunset. In situations where there is no motion within the scene every vertical line is identical, and the repeated line image comprises unbroken horizontal bright and dark bars. If a large object were to pass through the scene the brightness levels in several lines would be noticeably changed. The duration of the disruption would depend on the velocity of the object but, in all cases, the original shape of the object would be radically distorted (Figures 13, 16, and 17).

Figure 153 shows an exotic example of the motion in the Martian scene. As the shadow of the meteorology boom moves from left to right, repeated scans through a single vertical line create a pseudo image. Note that this is the mirror image of the boom shadow as it actually is projected on the Martian surface at any instant (Figure 152).

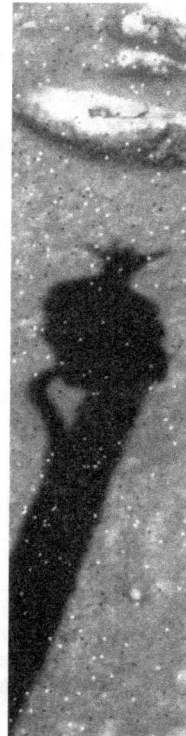

Figure 150   Figure 151   Figure 152   Figure 153

Early in the mission we used single line scan repeatedly, primarily to search for movement of wind driven sand which would have been recorded as irregular speckling throughout the line scan image. The results have been less than dramatic. No moving objects, large Martians or small sand grains, have been detected within the scene.

Digging In

Figure 154

Figure 155

Figure 156

Acquisition of soil samples for analysis within the Lander was the single most demanding Viking activity, requiring the integrated contributions of many scientists and engineers. In the course of the mission the surface sampler arm was deployed many times; numerous trenches were dug and many samples were delivered to instruments within the spacecraft. Figures 154 and 155, which show the extended sampler arm on Viking 2 from the vantage point of both cameras, capture the essence of a sampling operation. Figure 156 shows the same region after several trenches have been dug and one rock pushed a short distance. The trenches are visible at the far left and far right. The displaced rock, nicknamed Mr. Badger after a character in Wind in the Willows, is near the middle of the picture. Mr. Badger is seen in its original position in Figure 155.

We were fortunate that Martian surface materials were unconsolidated and relatively easy to shovel. Figures 157 through 161 show the progressive excavation of an area at the Lander 1 site known as Sandy Flats. Over a period of 10 months, more than a dozen individual trenching operations at Sandy Flats were completed. After it had been determined that the upper soil was biologically sterile, some scientists speculated that organisms might exist deeper in the soil where they would be protected from deadly ultraviolet radiation. The thesis was tested by repeatedly digging at the same spot until a hole more than 15 cm in depth was excavated (Figure 161). In one sense this was a sophisticated engineering task, requiring complicated sequences of sampler arm commands. In another sense it was reminiscent of childhood exercises with sand and shovel. At any rate, it was to no avail. Samples yielded negative biological results.

Figures 162 through 164 record sampling of a duricrust layer at the Lander 2 site. Figures 162 and 163 are two views of the area before sampling. Because this small exposure of duricrust was nestled between large blocks, engineers were concerned that the sampler arm might be damaged by inadvertently striking one of the boulders. Stereoscopic analysis of picture pairs indicated that the duricrust was accessible, but that the arm would come very close to several potentially disabling obstacles. Our nervousness was dispelled only when we received a picture showing the virtual obliteration of the duricrust patch (Figure 164).

Figure 157

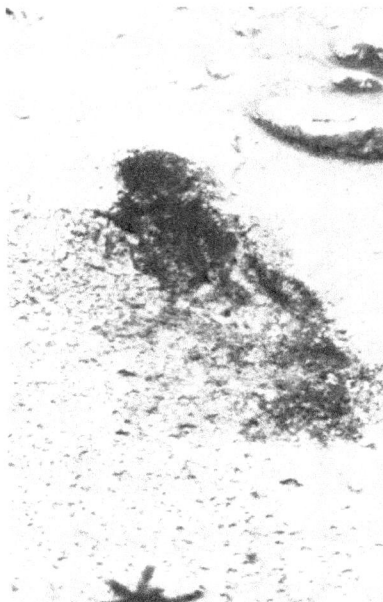

Figure 158

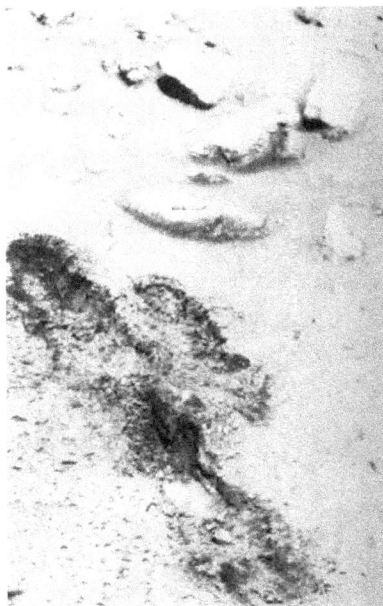

Figure 159

Figure 160

In the search for Martian organisms it was suggested that material from under rocks might be an attractive environment for organisms (protected from the sterilizing effects of the solar ultraviolet). Figures 165 through 167 show a boulder (nicknamed "Notch Rock") at the Lander 2 site that was successfully nudged to one side. Figure 165 is the "before" picture. Figure 166 shows the sampler arm midway through its pushing operation. Figure 167 shows the completed activity. The rock has been moved approximately 3 cm. A small depression is visible in front of the rock where it had previously been partially buried by surface material. Again, no organisms were found.

Figure 161

The decision to push a rock with the relatively fragile sampler arm was preceded by many days of careful analysis and consideration of potential risks. During a long meeting at which we discussed the relative merits of several candidate rocks, a foreign scientist, visit-ing JPL, sat briefly in the back of the room. Already impressed with the exotic space age technology in evidence throughout the laboratory, a puzz-led frown crossed his face as he listened to a group voting for a favorite rock.

Figure 162

Figure 163

Figure 164

Figure 165

Figure 166

Figure 167

Secondhand Spacecraft

In the course of the Viking mission the Landers, carefully cleaned and sterilized before leaving Earth, slowly accumulated a layer of dirt after arrival on Mars. Some of this was soil that was spilled on the upper deck when the surface sampler dumped sediment into the three funnels leading to the biology, GCMS, and inorganic chemistry instruments.

A grid had been painted on the upper deck of the Lander especially to create a distinctive background against which accumulation and movement of dust could be measured (Figures 168 and 169). In order to more clearly map the distribution of dust, parts of these two images have been rectified as though the viewer were looking straight down on the Lander grid (Figures 170 and 171). This also allows easy comparison between images acquired by the two cameras.

We recognized early the sediment spilled from sampling operations and some purposely dumped on the deck for analysis of soil properties. However, it was several months before we realized that the swirls of yellowish sediment were changing when no surface sampler activity had occurred.

Figure 168   Figure 169

The changing sediment patterns are caused by the Martian wind. Why is sediment movement observed on the Lander but not in the natural scene? Several possible reasons come to mind. Surface materials may have been disaggregated during transportation to the spacecraft deck, creating finer grains easier to move. Constriction of air flow by spacecraft components may increase wind speed and turbulence. Vibration of the upper deck may cause particles to bounce, thereby becoming more susceptible to lateral transport by wind. Finally, a similar amount of dust transport may be occurring on the Martian surface, undetectable in the absence of a background as contrasting in color as the Lander top.

During the extended mission giant dust storms occurred in the southern hemisphere and spread considerable material into the northern hemisphere.

Although wind velocities at both Viking sites remained below the critical velocity necessary for initiation of sediment transport, extensive dust clouds darkened the sky. There is an indication that fine material has been settling out from those clouds, progressively covering the spacecraft with a thin layer of fine dust. We also suspect that some dust has been redistributed throughout the Viking 1 site.

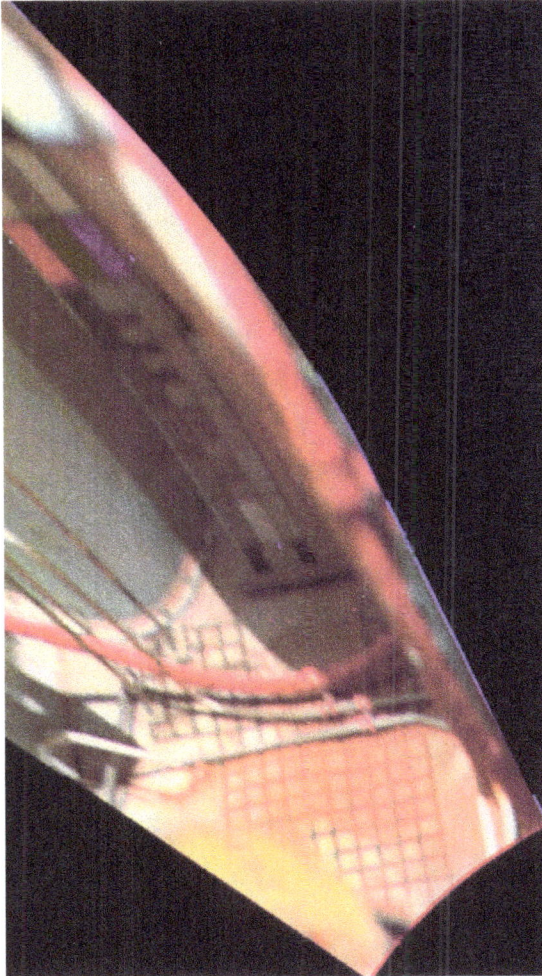

Figure 170

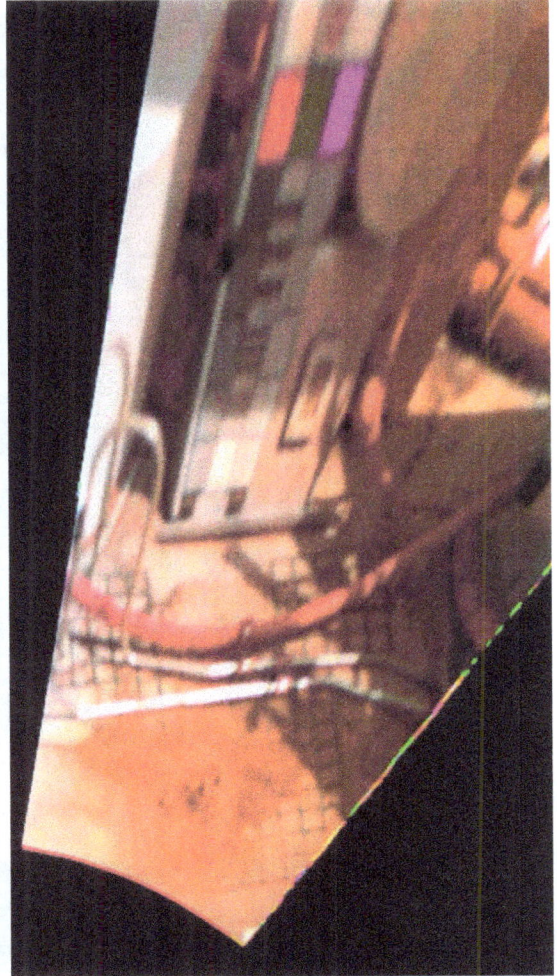

Figure 171

The Changing Atmosphere

As previously mentioned, dust storms occurred in the southern hemisphere during the extended mission, creating large dust clouds which eventually covered much of the planet. In addition, as the northern winter approached and temperatures declined, ice condensed on atmospheric dust particles. These two effects, which could change over a few hours, conspired to make prediction of apparent Sun brightness very difficult. We encountered a problem well known to all photographers: trying to guess light levels. Our problem was unusually difficult, however, because we had to prepare the camera commands a week or more in advance of the actual picture. Sometimes we guessed wrong and pictures were either overexposed or underexposed.

Figure 172, taken during the extended mission, is generally darker and shadows are more diffuse than equivalent images taken during the early mission. This is the result of increased atmospheric opacity and scattering of light by suspended dust particles. Note, in particular, the "soft" circular shadow of the S band antenna.

Figures 173, 174, and 175 record variations in apparent brightness of the Sun, imaged directly with the Sun diode. Because only the Sun is bright enough to be recorded, the pictures are relatively unimpressive, although scientifically useful. Figure 173 shows the Sun in a relatively clear sky. Figure 174 shows a darkening by a factor of more than 2, the Sun actually being invisible. Figure 175 shows a partial return to clear conditions.

Figure 172

Figure 173   Figure 174   Figure 175

Beyond Mars

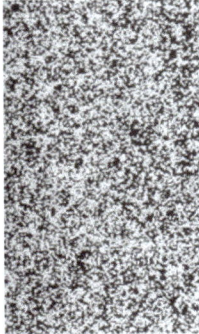

Figure 176

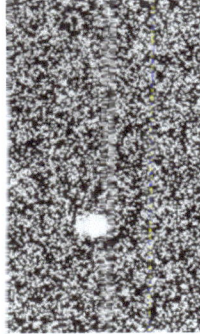

Figure 177

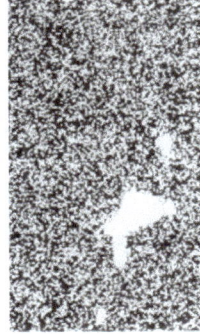

Figure 178

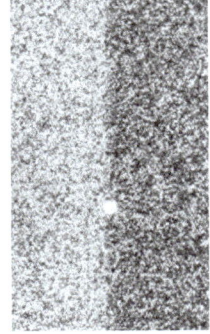

Figure 179

There is an undeniable fascination in photographing the familiar objects of the solar system from an unfamiliar vantage point. The best examples, of course, are the photographs of Earth from deep space.

We had hoped to photograph Earth from the surface of Mars, even though we knew it would appear only as a pinpoint of light. Unfortunately, we discovered that our camera lacked the sensitivity to discern such a faint object. The problem was intensified by the fact that Earth always appears close to the Sun, just as Venus does when viewed from Earth. This is the general situation when viewing a planet with orbital radius smaller than that of your own planet.

We were restricted to viewing two objects beyond Mars, Phobos, and the Sun. Figures 176 through 178 show the Sun imaged at different elevation angles. As it rises higher and the path length of light through the atmosphere decreases, the solar disc becomes brighter. Figure 176 is taken at 5:48 a.m. The Sun, about 5° above the horizon, is barely distinguishable. The speckled appearance of the sky is an artificial effect, caused by extreme enhancement of detail in a very bland scene. Figure 177 was taken at 6:12 a.m.; the elevation angle of the Sun is about 10°. Figure 178 was taken at 7:00 a.m.; the Sun is approximately 20° above the horizon. The irregular shape and secondary bright spots are due to spurious reflections of light from the outer camera housing.

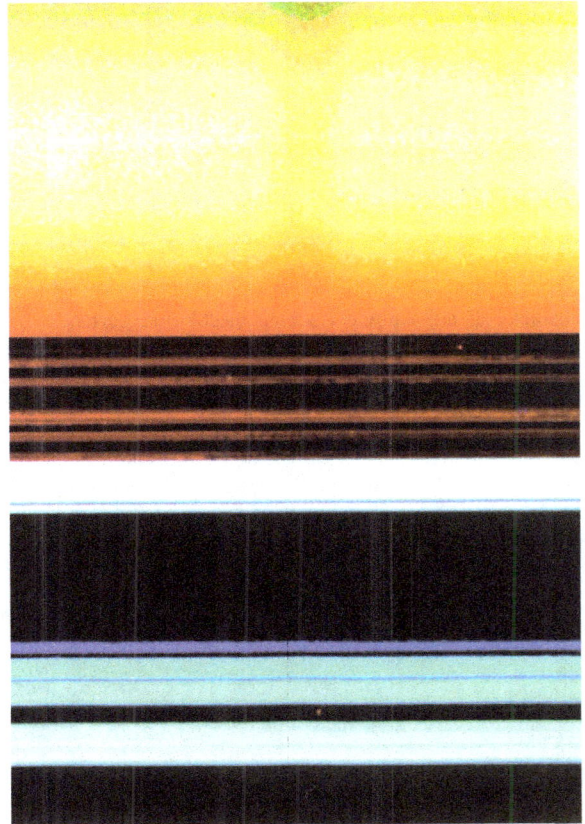

Figure 180

Phobos is one of two natural satellites of Mars. An irregular, cratered object about 22 km in diameter, it circles Mars approximately 1000 km above the surface every 11 hours. Although visible as no more than a bright object several pixels in diameter (Figure 179), images of Phobos proved very valuable in determining the opacity of the night sky. Color and IR reflectance values of the integrated disc were clues to the satellite's chemical composition. The spectral data best fit that of carbonaceous chondrite, a variety of meteorite believed to represent primitive solar system material.

Possibly the most unusual of all Viking Lander pictures records the passage of Phobos' shadow during a solar eclipse (Figure 180). On Earth the apparent size of the Moon is exactly the same as that of the much larger but more distant Sun. Consequently, direct sunlight is completely blocked out during a total solar eclipse. On Mars, Phobos covers only a quarter of the solar disc. However, passage of the penumbral shadow causes a general drop in light level that is instrumentally detectable. Figure 180 is a repeated line scan image looking back across Viking Lander 1. The colors have been distorted purposely to enhance detail. The blue and white horizontal stripes correspond to test chart patches. The brownish stripes in the middle represent the Martian surface visible above the spacecraft. Note a decrease in light levels in the sky midway through the imaging event. The darkening, caused by the passage of the penumbral shadow of Phobos, is present over approximately 100 vertical lines. Moving at about 2 km sec, the shadow took about 20 sec to pass over the Lander. It should be obvious that successful acquisition of this picture requires extremely accurate calculations regarding the orbit of Phobos about Mars. Not only was the passage of the small shadow across the Lander correctly calculated, the time of the event also was predicted within a few seconds.

Coloring Mars

As we have already discussed in the introductory text, the problems in reconstructing the colors of the Martian surface and atmosphere are formidable. It would be nice to present a folio of color pictures with the unqualified comment that these are the colors of Mars. But, after more than a year of analysis, it becomes clear that the situation will never become so simple.

In order to correct for the out of band spectral leaks in the color filters within the camera it is necessary to have access to both a color and an infrared (IR) picture taken at the same time. Only in that way can the IR contributions to the visible color, a result of camera filter design, be assessed. For the many color pictures taken without an IR companion (e.g., Figures 186, 192, and 198) detailed calibration is not possible. Instead, the three visible channels – blue, green, and red – are balanced on the assumption that there are no out of band contributions.

Even for pictures where we have both visible and IR information, two possible types of color can be created. The character of the sunlight reaching the Martian surface is significantly different than that reaching Earth's surface, primarily because of scattering and absorption by suspended dust particles. In addition, a certain amount of light reflected from the yellowish brown surface finds its way back as reflected sky light.

Figures 181, 182, and 183 illustrate the range of possible color reconstructions. All three pictures are based on the same camera data. The sampling area at the Viking 1 site is shown. Two trenches in the Sandy Flats site, the first to be dug are shown at the far left. Figure 181 is produced by using only visible color information, making no allowance for IR leaks. The scene has a reddish or orangish cast. Figure 182 is constructed by incorporating data from an

Figure 181

accompanying IR picture and using the spacecraft's test charts, in this way accounting for out of band contributions and atmospheric colorations. The scene is shown as it might appear "on Earth." For example, if you could pick up one of the boulders and transport it to your back yard, this is the color you would observe. Note that, relative to Figure 181, the reddish tint is subdued, replaced with a brown hue. Figure 183 shows the scene as it would appear "on Mars." The yellowish cast of sunlight filtering through the atmospheric dust imparts a similar yellowish hue to the entire scene.

Figure 182

Figure 183

Figure 184 is a view looking across Lander 1. Two test charts are visible. The one at the left is directly illuminated; the colors of the red, green, and blue color chips are accurately rendered. The test chart to the right is in shadow and, consequently, of little value for color calibration.

Figure 185 shows the landscape to the left of the region in Figures 181 to 183. Drifts of sediment are visible in the distance. Footpad 2, at the bottom of the image, is buried by yellowish brown soil.

Figure 186 is a historic picture, the first color image to be taken on the surface of Mars. Because there was no accompanying IR image, it is not possible to compensate for irregularities in the camera color filters. The color has the same relative quality as in Figure 181

Figure 184

Figure 185

Figure 186

Figure 187

Figure 188

Figure 187 (as on Mars) and Figure 188 (as on Earth) show the landscape in the vicinity of the large dark rock, "Big Joe." Note that the boulder has a crown of fine grained material. Exactly how this mantle formed is a puzzle. It may be an erosional remnant of a dust layer which formerly covered the region with a thickness of a meter or more.

Figure 189   Figure 190   Figure 191

Figure 193

Figure 192

Figure 194

Figure 195

Figure 192, taken early in the Viking 1 mission, was nicknamed "The American Flag Picture" even before it was taken. During the final months of preparation before the landing NASA managers took exception to our decision to feature only the Mars surface in the first few color pictures. In particular, they emphasized the popular appeal of a color picture looking back across the spacecraft the American flag in the foreground and the Martian horizon in the distance. We temporized; they insisted   Figure 192 was the result.   The photograph is both pictorially attractive and scientifically useful. The bright ridge in the distance, part of the crater rim described in Figures 77, 78, and 79. catches the morning Sun. Difficulties in precisely balancing the colors are indicated by the violet hue of the blue color chip on the test chart. The relative differences in reflectance that yield a color image are illustrated by the three constituent images taken in blue light (Figure 189), green light (Figure 190), and red light (Figure 191).

Figure 193 is a view looking back across Lander 2. Dark boulders are prominent against the yellowish brown soil, standing like regularly deployed sentinels out to the far horizon.

In the course of the Martian year an important surface change was observed at the Viking 2 site. During the winter months a thin layer of frost formed. This is documented by comparing two black and white pictures, the first taken early in the mission in late Martian summer (Figure 194), and the second taken months later in late Martian winter (Figure 195). Patches of receding frost are visible in the latter picture. During the height of winter, frost probably covered the entire surface. Figure 196 is a color image of the frost, looking across the spacecraft. The colors purposely have been distorted slightly to enhance color contrast.

Figure 196

The frost persists at temperatures higher than those required for frozen carbon dioxide (dry ice). It may be water ice or a carbon dioxide clathrate, a crystalline mixture of H_2O and CO_2.

The region in front of Lander 2 was photographed many times in color. In Figure 197 the surface sampler shroud contrasts sharply with the remainder of the scene. Dust clings to the surface of footpad 3 and the lower part of the leg strut.

Figure 198 illustrates the operational versatility of the Viking cameras. By generating special commands we were able to combine operational modes in ways never envisioned by the original designers. In this picture the 0.12° color diodes were stepped at 0.04° intervals instead of the usual 0.12° intervals. The result is a "high resolution" color image in which spatial resolution approaches that of a black and white picture for which both aperture and stepping angle are 0.04°. The red and green patches near the rock in the upper right occur because the three blue, green, and red images that comprise the color picture were taken on successive days. In the intervening period the sampling shovel disturbed the soil. False colors result primarily from differences in shadow distribution before and after the sampling event.

Figure 199 can best be assigned to the special effects category. It illustrates one of the many ways in which black and white images can be combined and printed in color to yield unusual patterns. In this instance an early morning image is assigned the color yellow and a picture taken in the afternoon is assigned the color blue. Although colors produced in this way are completely false, important details in the scene are sometimes favorably accentuated.

Figures 200 and 202 are a related pair, nicknamed the "sundown" and "sunset" pictures. Early in the mission, when it became clear that there was a large amount of scattering particles in the lower atmosphere, we realized the potentially dramatic appeal of low Sun pictures. However, the first priority was to document the landscape with high resolution and color images. Following several weeks of successful operation on the Martian surface we relaxed to the point of attempting exotic but high risk pictures. The sundown picture was taken on sol 30, just 5 min before the sunset picture. The Sun is to the right, only 4° above the horizon. When the sunset picture was taken the Sun was 2° below the local horizon. The banding in the sky is an artifact produced by the camera. The brightness actually changes continuously, moving away from the Sun, but the natural continuum is divided into incremental brightness levels by camera electronics. If the number of quantified brightness levels were much larger, then the closely spaced steps would more closely approximate the continuous gradation of the natural scene. The slightly different color from band to band is also an artifact, caused by the fact that the brightness boundaries for the three color channels are not coincident. The general tendency for the sky to become bluer in the direction of the Sun is real, caused by dominant short wave length forward scattering of light in that direction. No planets or stars are visible in the evening sky. The occasional pinpoints of color, especially noticeable in Figure 203, are artifacts caused by noise in data transmission and by mismatching during image enhancement.

Figure 197

Figure 198

Figure 199

Figure 200

Initially dismissed as little more than oddity, the sunset picture has been reproduced many times, especially in its more exotic "stretched" version (Figure 203). Admittedly, this version is more an art object than a scientific product. The colors have been specially enhanced to bring out the detail in the foreground. However, the visual impact of this computer painted scene is universal: the Martian twilight envelopes the viewer.

Later in the mission the sunset picture was matched with a sunrise picture at the Viking 2 site (Figure 201) Again, the Sun is several degrees below the horizon. The relatively pronounced blue shading in the sky is probably a result of early morning ice crystal fog.

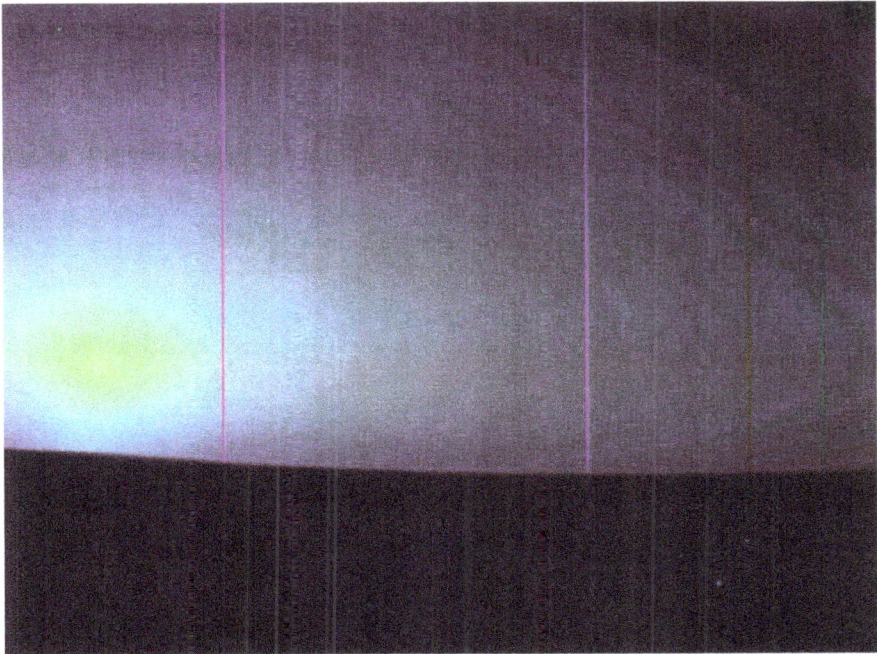

Figure 201

Figure 202

Figure 203

A Different Perspective

Your impression of Mars depends upon your point of view, literally. Using an extreme example, the landscape as depicted by the Lander cameras, situated a little more than 1 m above the surface, has little in common with the landscape photographed by the Orbiter cameras, 1500 km above the surface. The principal difference is in scale. The resolution from orbit is about 50 m, precluding any possibility of directly detecting the Lander. In addition there are few, if any, large diagnostic land forms that can be identified in both Orbiter and Lander pictures. This situation may be improved in the future. Newly obtained higher resolution Orbiter images will allow us to refine the position of the Lander to within about 40 m.

Figure 204 illustrates our best estimate at correlation of Lander and Orbiter pictures. Lander 1 panoramas have been computer distorted into a circular format, something like the image obtained with the fisheye lens of a conventional film camera. This makes it easier to correlate the image with an Orbiter picture. For example, moving out along a straight line from the center, any feature intersected in the Orbiter picture should also be intersected in the Lander picture. The crater rim indicated at point A in the Lander picture is probably the same one indicated at point B in the Orbiter picture. (For a better rendition of the crater, see Figures 77, 78, and 79.) Other nearby distinctive features in the Orbiter picture may be masked in Lander pictures by local hills and rises. Incidentally, the landing site viewed from orbit resembles a sparsely cratered lava plain, very similar to the dark lunar maria. This is consistent with the interpretation from Lander pictures but, at this more detailed level, the lavas are seen to be broken up and modified by wind action and by other processes.

Viking 2 landed on a surface that is difficult to interpret from Orbiter pictures. A muted, mottled texture suggests covering up of subjacent, sharper topography. According to one interpretation, we landed on the outer part of a vast ejecta deposit associated with the 100 km crater, Mie, located 160 km to the northeast. The Lander pictures neither

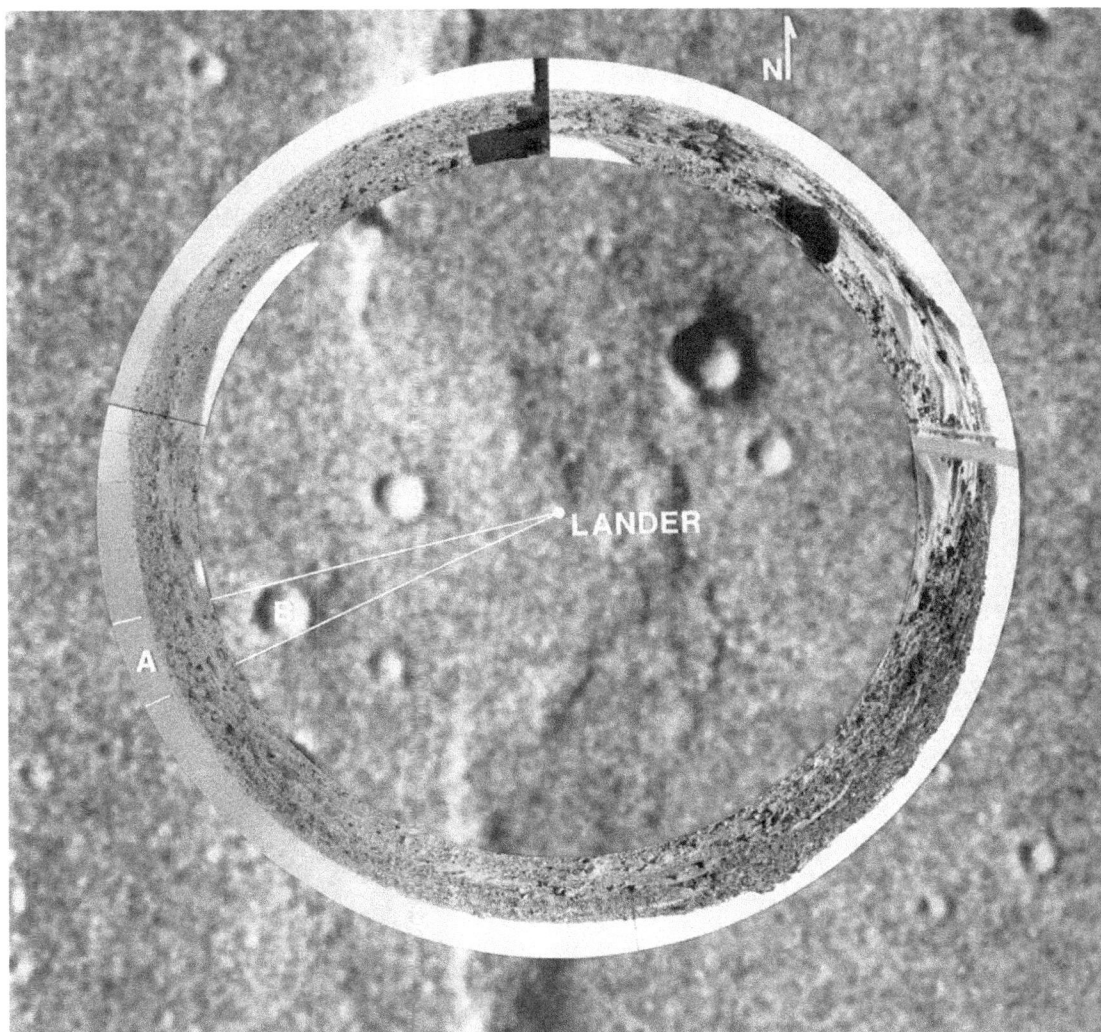

Figure 204

rule out nor confirm this hypothesis. The large blocks could be the residue of a former thick blanket of ejecta that has been winnowed by the wind. No land forms have been uniquely correlated in Orbiter and Lander 2 images.

Viking 1 landed with a tilt of 3° down toward the west. The tilt of Viking 2 is much more noticeable, 8.5° down toward the west (Figure 205). In both cases the tilt causes the horizon to appear as a gentle, undulating curve over the azimuth range of 360°. When the Lander 2 scene is rectified to the perspective of a vertically oriented camera, it appears as in Figure 206. Note that the horizon is now almost perfectly horizontal.

Because of the way the camera scans, considerable distortion is introduced in images taken close to the Lander at large depression angles. Individual scan lines can be compared to lines of longitude on a globe. Close to the equator, the spherical surface can be represented on flat paper with little distortion. But, as the lines of longitude approach the pole, they begin to converge. If they are shown on a two dimensional map with the same spacing as at the equator, land masses near the pole, such as Greenland or Antarctica, are greatly distended. This same effect causes objects photographed near 0° to be faithfully rendered, but objects at the lowest elevation angle of 60° are elongated by a factor of two. The magnitude of the effect can be calculated by comparing Figure 207 with Figure 24. In the latter case the usual camera perspective is altered to a planar perspective, similar to what your eye would perceive looking directly down on the Martian surface.

Figure 205

Figure 206

Figure 207

The Third Dimension

The Viking Landers have two eyes and therefore can view the landscape in three dimensions, just as we do with our own eyes. However, presentation of stereoscopic pairs has been a frustrating problem. Because the camera separation is almost 1 m, objects close to the spacecraft are viewed with orientation equivalent to that for an object at the tip of one's nose. The pictures have to be carefully rectified so that the relative scales remain the same throughout the scene. The distance between the images must be controlled to permit optimum fusion.

Some persons can appreciate the stereoscopic effect much more easily than others. There is, however, one fairly reliable guide to the viewer's success. If as he peers through the stereoscope, you ask him if he sees the third dimension and he responds noncommittally "yes," then you know he has not. Wait a few minutes and you will hear an exclamation of surprise and wonder. Then you will know he has seen it. The effect is so unusual, literally drawing you into the scene, that very few people come upon it without excitement.

On the following pages we present 19 stereo pairs. To look at them use the viewer enclosed in the envelope inside the back cover. Most persons find it easiest to focus their attention on some prominent object in the farfield, preferably on the horizon. Position the centerline of the viewer equidistant between the two images of the object. Looking through the viewer, fuse the two images into a single image. To assure yourself of a fused image rotate or translate the viewer slightly. You will see the two images separate and then fuse as the viewer is returned to proper orientation.

For some persons, fusion will not be accompanied immediately by a three-dimensional effect. If you are uncertain regarding your own success, continue to look at the landscape, moving the viewer slightly if necessary. When the third dimension is optimally perceived, a succession of ridges and depressions will be so sharply delineated as to give the impression of a series of discrete vertical planes, moving away from the spacecraft. When you have obtained the appropriate effect for the farfield, work your way back into the nearfield. Once you "lock in," subsequent attempts will be considerably easier. Perhaps the most striking phenomenon to be observed stereoscopically is the undulating character of the scene, particularly at the Viking 1 site. Figures 208 to 215 comprise a panorama in front of the spacecraft, moving in a clockwise direction. Figures 216 and 217 reveal details of rocks in the foreground. Figures 218 and 219 show topography in the vicinity of the sediment drifts. Figures 220 to 223 comprise a second panorama looking to the right of the Lander toward the northwest. Drifts of sediment, intermixed with clusters of boulders, occur throughout the scene.

Three stereo pairs, Figures 224 to 226, have superposed on them contour lines generated with the aid of a computer. In one case (Figure 226) the lines are drawn to represent horizontal planes intersecting with the surface, similar to a conventional topographic map. In the other two images (Figures 224 and 225) the planes are vertical so that vertical profiles are superposed. During the mission, this combination of graphic and pictorial data proved a valuable aid in planning movement of the surface sampler arm.

The last two Figures, 225 and 226, are views of the Lander 2 site. Although there is some relief in the scene, the topographic variations are not as dramatic as at the Viking 1 site.

This is an appropriate place to end our discussion of Mars. So vivid is this three dimensional effect that, as you look at these ridges and depressions, you may get the impression you can actually step from your chair and onto the surface of Mars. Someday our descendants will.

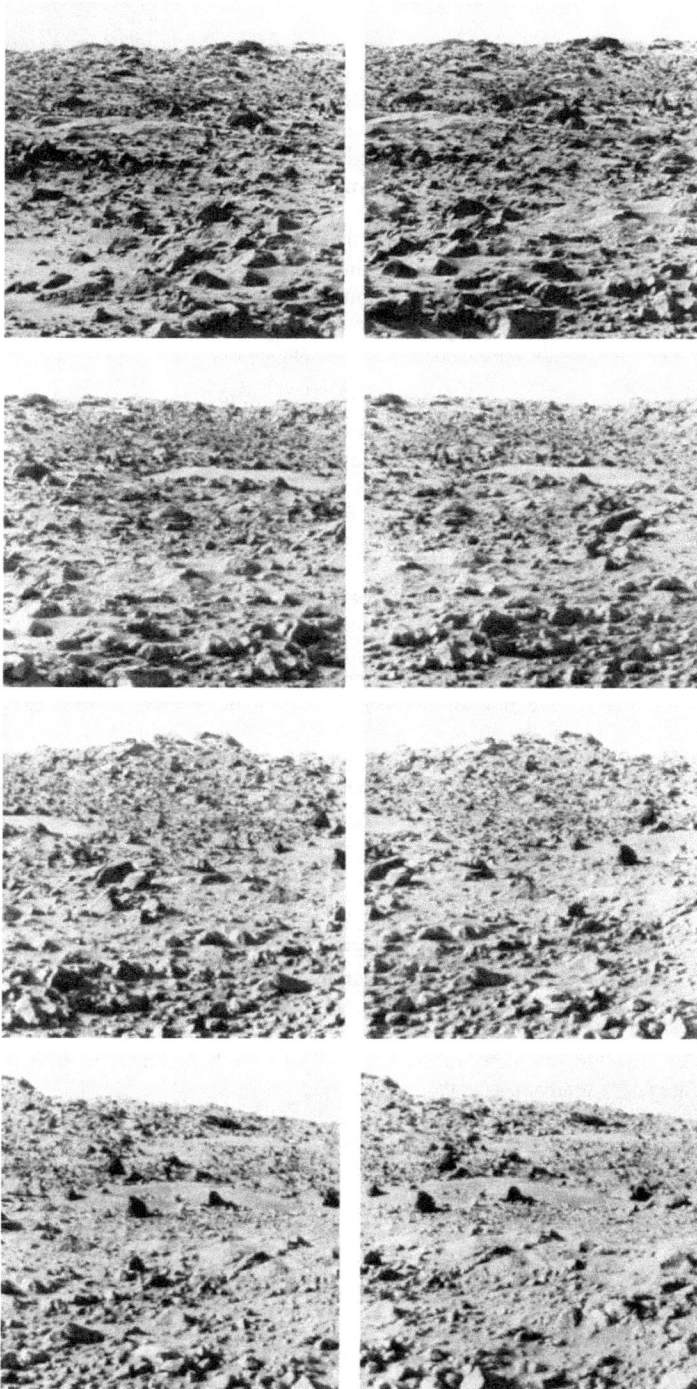

Figures 208 (top) - 211 (bottom)

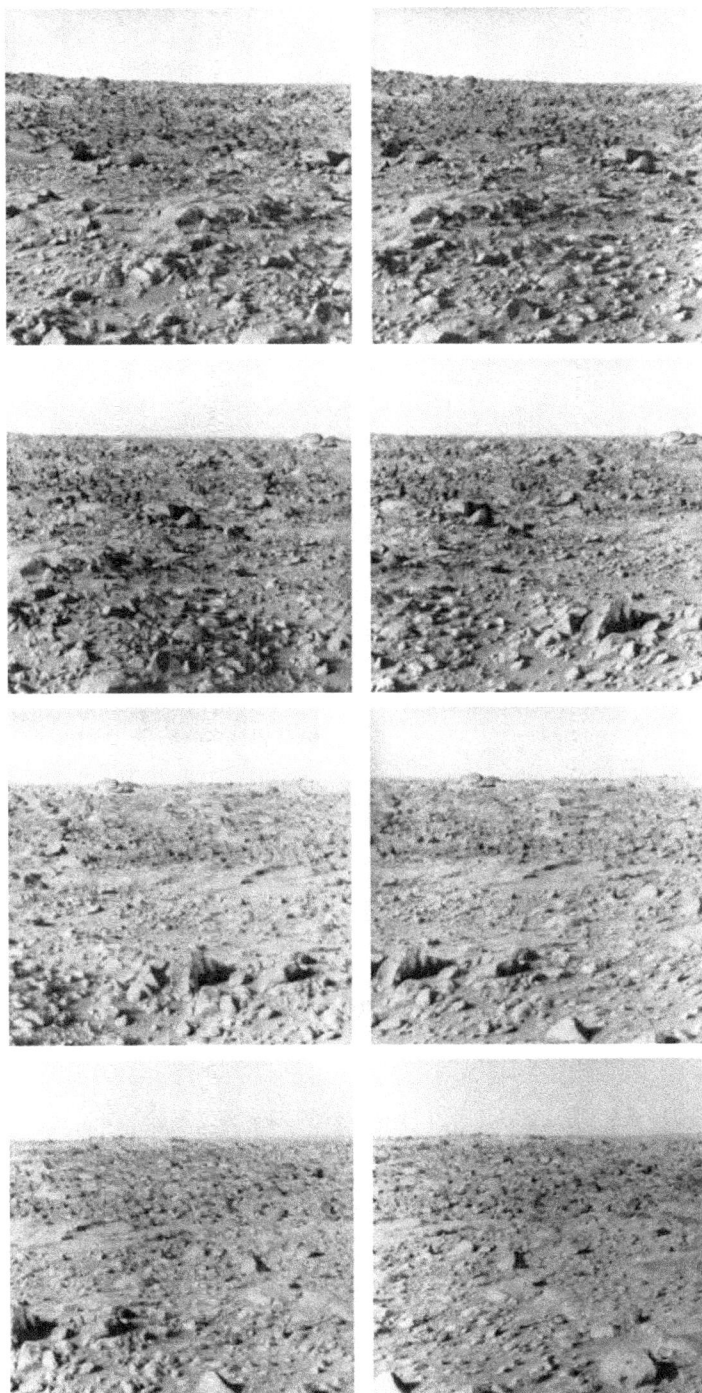

Figures 212 (top) - 215 (bottom)

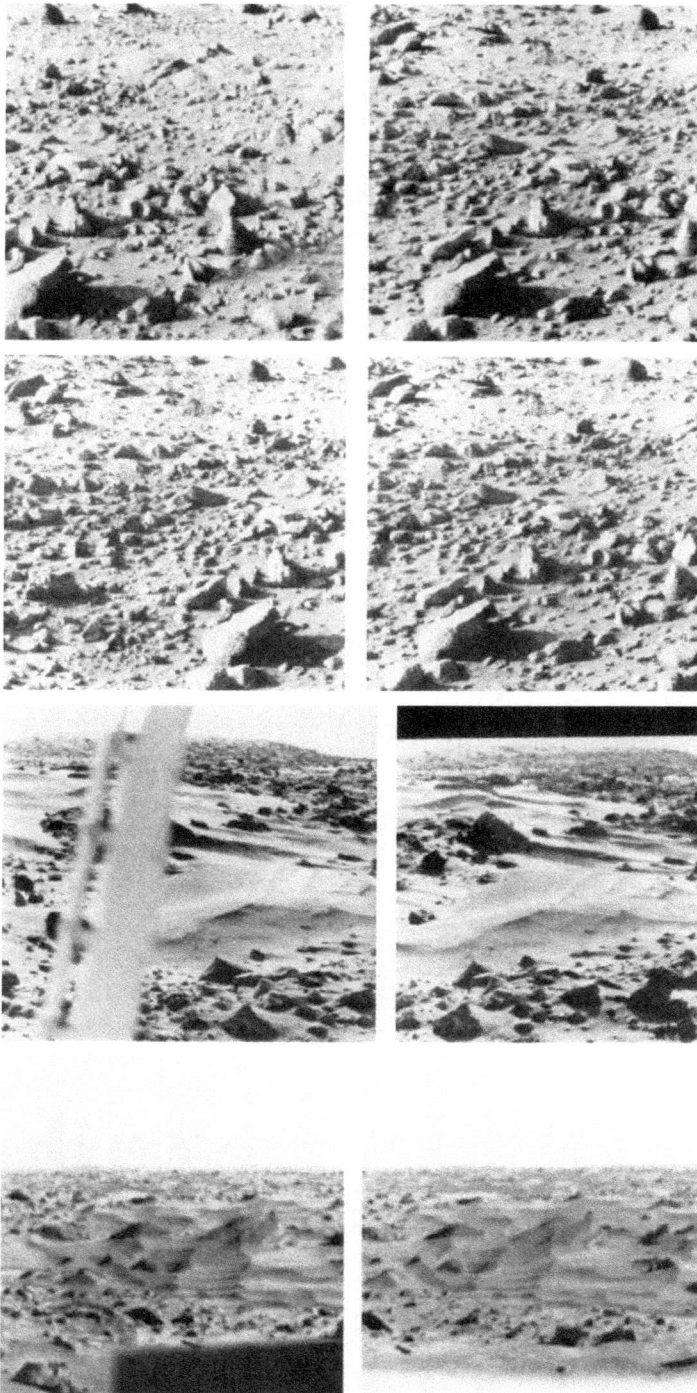

Figures 216 (top) - 219 (bottom)

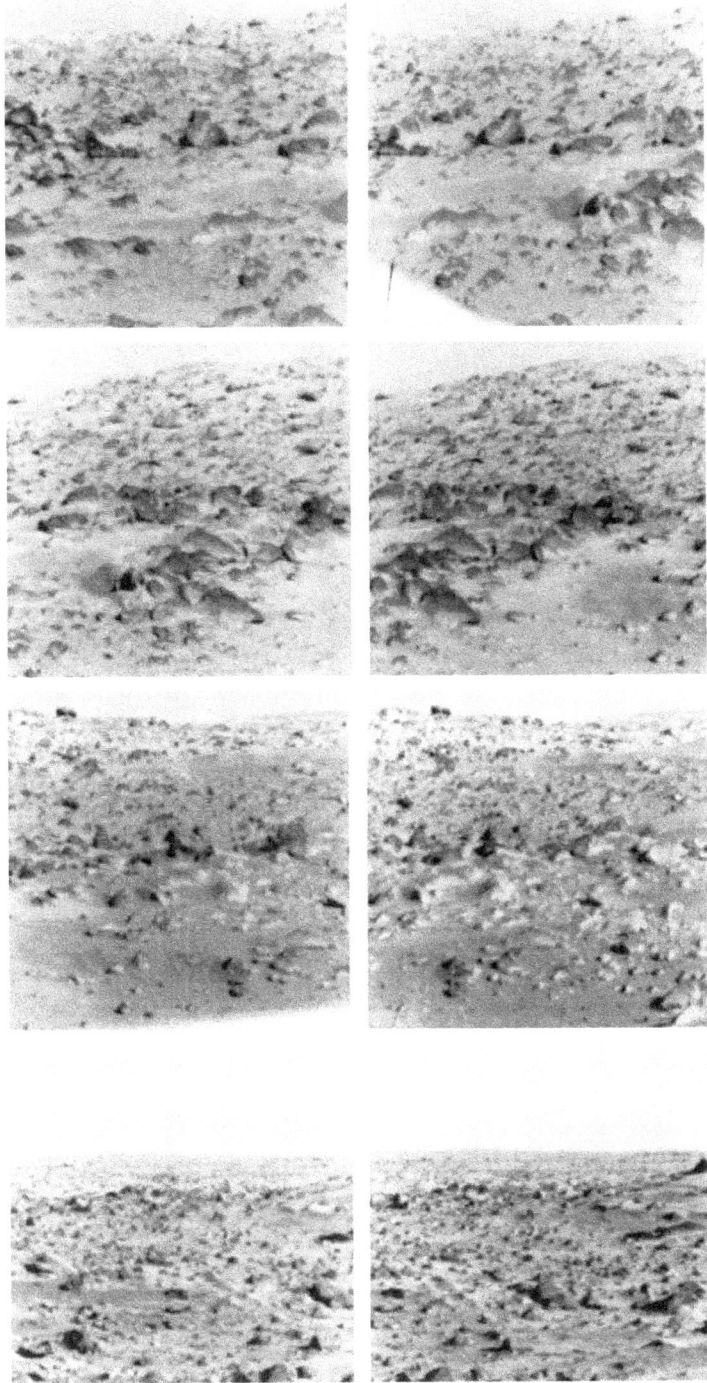

Figures 220 (top) - 223 (bottom)

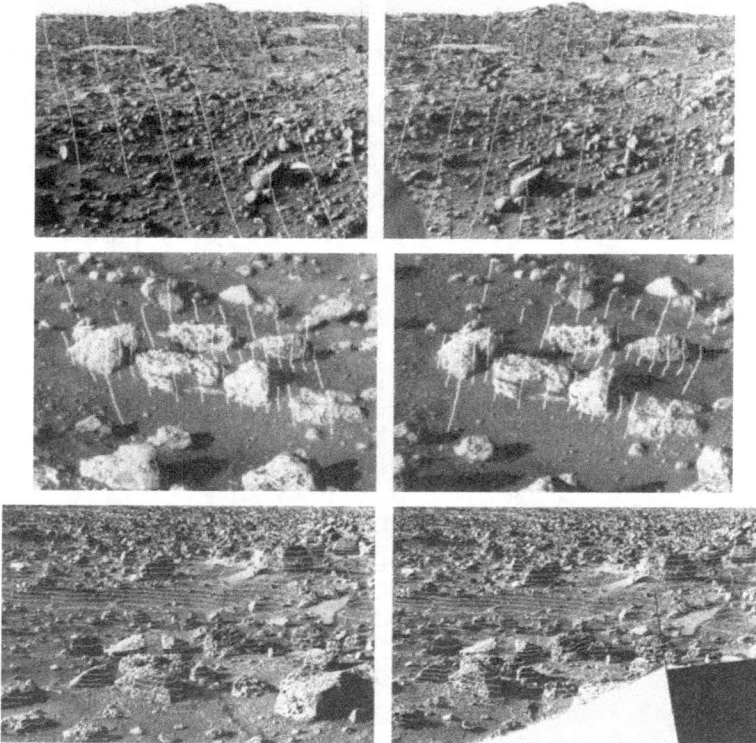

Figures 224 (top) - 226 (bottom)

Panoramic Sketches

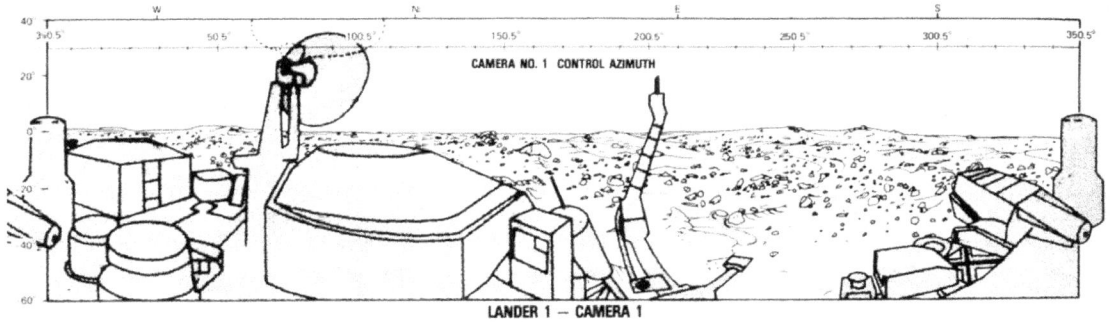

LANDER 1 — CAMERA 1

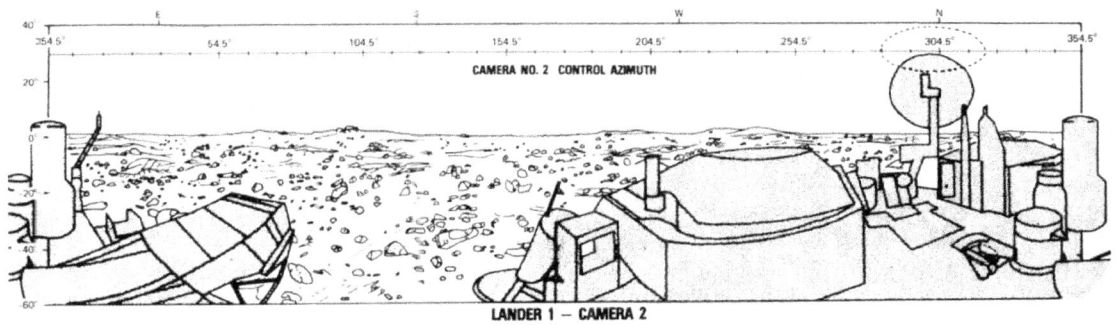

LANDER 1 — CAMERA 2

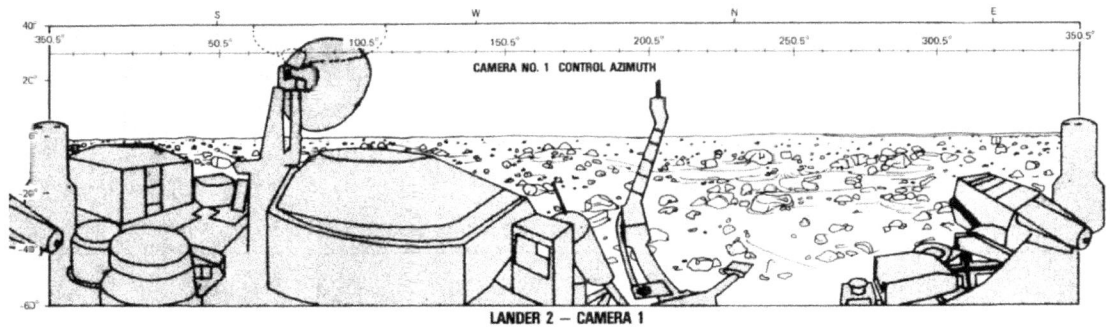

LANDER 2 — CAMERA 1

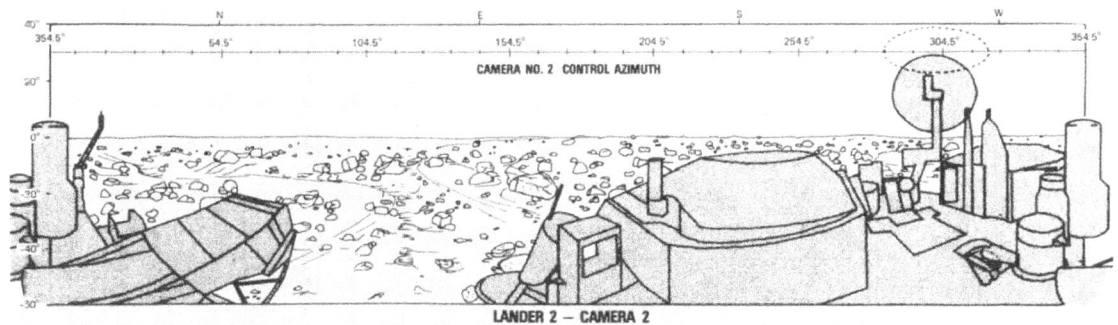

LANDER 2 — CAMERA 2

A Picture Catalog

This table is included for readers who wish to know the "camera settings" for particular pictures. The figure number in column 1 is followed by the camera event label in column 2. The first number in the camera event label indicates the Lander and the second number indicates the camera. The letter followed by three numbers indicates the picture number, indexed successively starting with the first picture after touchdown. The A series extends from A001 to A255 and is followed by the B series. Accordingly, the camera event immediately following A255 is B000. The first picture taken after touchdown on July 20, 1976, is labeled 12A001, indicating Lander 1, camera 2, series A, camera event 001.

Azimuth start and stop angles indicate the limits of the published figures. In many cases the total image corresponding to the camera event has been cropped. The camera control azimuths on the panoramic sketches (see preceding page) indicate the relative positions of particular azimuths. Also shown are compass directions relative to north.

Lower and upper elevation limits also refer to the published figures, and not necessarily to the limits in the original images. Refer to the accompanying skyline drawings for the relative positions of particular elevation angles.

Diode selections include low resolution black and white survey (SURV), color (CLR) which is actually a combination of blue, green, and red diodes, individual color diodes reproduced in black and white labeled BLU, GRN, or RED, Sun, and high resolution black and white (BB). In the latter there are four focus settings — 1, 2, 3, and 4 — corresponding to ranges of 1.9, 2.7, 4.5, and 13.3 m.

The significance of gain and offset is described in the introductory text. Most pictures were taken with a gain of 4 and an offset of 1, camera settings appropriate for scenes with relatively high light levels and considerable contrast.

Greenwich Mean Time (GMT) indicates the time (day, hour, minute) that the picture was taken, measured at Greenwich, England. Measurements at this locality are the basis of standard time throughout the world.

Local Lander time (LLT) indicates the time that the picture was taken, measured in terms of a Martian system. The first three numbers refer to the Martian day, or sol, starting with the day of landing, sol 000. This is followed by the hour and minute, measured after local Martian midnight. Lander 2 sols are not the same as Lander 1 sols, the former being offset in time because of a later landing.

The Sun azimuth at the time the picture was taken is measured clockwise from north, and the Sun elevation is measured above the local nominal horizon.

Figure Number	Camera Event Label	Azimuth start	Azimuth stop	Elevation top	Elevation bottom	Diode	Offset	Gain	Greenwich Mean Time	Local Lander Time	Sun azimuth	Sun elevation
24	12A001	102	160	-40	-56	BB1	1	4	202/11:53	000/16:13	285	39
25	12A136	102	137	-40	-56	BB1	1	4	226/01:58	023/15:07	280	52
26	12A002	10	110	+20	-40	SURV	1	4	202/11:59	000/16:19	285	37
27	11A018	120	235	+20	-40	SURV	1	4	205/12:11	003/14:32	281	61
28	11B061	130	230	+40	-20	BLU	1	4	240/10:11	037/14:06	276	66
29	11A251	139	169	0	-16	BB4	1	4	234/07:16	031/15:09	279	52
	11A143	169	185	0	-16	BB4	1	4	228/03:09	025/15:00	280	54

Figure Number	Camera Event Label	Azimuth start	Azimuth stop	Elevation top	Elevation bottom	Diode	Offset	Gain	Greenwich Mean Time	Local Lander Time	Sun azimuth	Sun elevation
30	11D095	150	130	0	-17	BB4	1	3	141/09:37	297/10:00	138	29
31	11A097	139	135	0	-20	BB3	1	4	216/12:24	014/07:30	73	27
32	11B097	139	185	0	-20	BB3	1	4	243/15:29	040/17:26	286	21
33	12A116	90	123	+10	-10	BB4	1	4	222/16:46	020/07:54	75	32
34	12A212	90	123	+10	-10	BB4	1	4	232/06 05	029/15:17	280	50
35	11A129	175	195	+5	-10	BB4	1	4	224/18:21	022/08:10	76	35
36	12D043	5	35	0	-13	BB3	1	4	126/02:49	082/13:05	189	46
37	12A010	77	118	0	-20	BB4	1	4	204/04:17	002/07:18	72	24
38	12A119	77	118	0	-20	BB3	1	4	223/00:07	020/15:15	281	50
39	11A017	272	300	-10	-30	BB3	1	4	205/12:07	003/14:28	280	62
40	11B052	252	286	-10	-30	BB3	1	4	239/02:50	036/07:25	75	25
41	11D039	105	140	-20	-40	BB2	1	4	126/01:44	282/12:00	169	43
42	12A164	175	204	0	-20	BB4	1	4	230/04:22	027/14:53	279	55
43	12A107	175	204	0	-20	BB3	1	4	220/15:02	018/07:30	75	27
44	12A003	83	101	-20	-33	BB2	1	4	203/04:07	001/07:48	73	30
45	12A140	83	101	-20	-33	BB2	1	4	227/02:34	024/15:04	280	53
46	11A078	220	241	-20	-38	BB2	1	4	214/12:00	012/08:25	76	38
47	11B144	220	241	-20	-38	BB2	1	4	263/01:45	059/15:09	275	50
48	12B112	282	305	-2	-17	BB3	1	3	244/16:21	041/17:38	287	19
49	12B016	282	301	0	-17	BB3	1	4	236/07:54	033/14:27	277	61
50	11A077	220	255	0	-20	BB4	1	4	213/18:22	011/15:27	282	48
51	11A077	235	250	+5	-5	BB4	1	4	213/18:22	011/15:27	282	48
52	12A010	103	136	0	-20	BB4	1	4	204/04:17	002/07:18	72	24
53	12A119	103	136	0	-20	BB3	1	4	223/00:07	020/15:15	281	50
54	12A116	126	165	+4	-10	BB4	1	4	222/16:46	020/07:54	75	32
55	12A211	120	160	-4	-10	BB4	1	4	232/05:34	029/14:46	279	57
56	12C158	125	125	-43	-60	BB1	1	4	074/18:04	232/13:20	205	43
57	11A022	265	265	-40	-60	BB1	1	4	206/05:30	004/07:12	71	23
58	11A079	182	235	-40	-60	BB1	1	4	214/12:10	012/08:35	77	41
59	11B050	182	235	-40	-60	BB1	1	4	238/09:37	035/14:51	278	55
	11A132	225	235	-40	-60	BB1	1	4	225/01:21	022/15:10	280	51

Figure Number	Camera Event Label	Azimuth start	Azimuth stop	Elevation top	Elevation bottom	Diode	Offset	Gain	Greenwich Mean Time	Local Lander Time	Sun azimuth	Sun elevation
60	12A152	130	180	-20	-40	BB2	1	4	229/03:46	026/14:56	279	54
61	12A152	160	180	-20	-32	BB2	1	4	229/03:46	026/14:56	279	54
62	12A108	142	168	-20	-40	BB2	1	4	220/15:12	018/07:40	74	29
63	12A078	155	190	-20	-32	BB2	1	4	214/12:00	012/08:25	76	38
64	12A002	110	310	+20	-40	SURV	1	4	202/11:59	000/16:19	285	37
65	12A018	10	140	+20	-40	SURV	1	4	205/12:11	003/14:32	281	61
66	11A129	100	160	0	-7	BB4	1	4	224/18:21	022/08:10	76	35
67	11A250	100	150	0	-7	BB4	1	4	234/06:46	031/14:38	278	59
68	12A235	150	210	0	-10	BB4	1	3	323/08:34	029/17:46	288	18
69	12A235	210	253	0	-10	BB4	1	3	232/08:34	029/17:46	288	18
70	12A125	228	260	+10	-8	BB4	1	4	223/17:41	021/08:10	76	35
71	11B073	35	60	0	-15	BB3	1	4	242/11:56	039/14:32	276	60
72	11B099	40	60	0	-20	BB3	1	3	243/15:47	040/17:44	287	17
73	11A111	40	60	0	-20	BB3	1	4	221/15:32	019/07:20	73	25
74	12A255	290	331	+10	-10	BB4	1	4	235/07:18	032/14:32	278	60
75	11A154	66	100	+8	-8	BB4	1	4	229/20:49	027/07:20	73	25
76	12B063	302	320	+8	-10	BB4	1	4	241/04:31	038/07:47	76	30
77	12A116	163	203	+5	-8	BB4	1	4	222/16:46	020/07:54	75	32
78	12A237	163	203	+5	-8	BB4	1	4	223/06:09	030/14:41	278	58
79	12A235	172	203	+1	-3	BB4	1	3	232/08:34	029/17:46	288	18
80	22A001	92	160	-40	-58	BB1	1	4	247/22:38	000/09:49	119	51
81	22A020	92	120	-40	-58	BB1	1	4	251/00:13	003/09:25	113	47
82	22A002	10	312	+20	-40	SURV	1	4	247/22:44	000/09:56	120	52
83	21B010	82	115	+9	-9	BB4	1	4	280/16:46	032/06:50	86	20
84	21B010	109	143	+9	-9	BBç	1	4	280/16:46	032/06:50	86	20
85	21A237	82	111	+10	-10	BB4	1	4	277/20:02	029/12:06	175	60
86	22A236	90	118	-2	-20	BB4	1	4	277/14:42	029/06:45	84	20
87	22A252	92	118	-2	-20	BB4	1	4	278/20:43	030/12:06	176	60
88	21A147	182	215	+8	-10	BB4	1	4	268/19:20	020/17:20	272	25
89	22A164	182	215	+8	-10	BB4	1	4	270/10:30	022/07:10	88	24

Figure Number	Camera Event Label	Azimuth start	Azimuth stop	Elevation top	Elevation bottom	Diode	Offset	Gain	Greenwich Mean Time	Local Lander Time	Sun azimuth	Sun elevation
90	22A178	98	128	-20	-40	BB2	1	4	271/11:09	023/07:10	88	24
91	22A206	127	158	-20	-40	BB3	1	4	273/12:25	025/07:07	88	24
92	22B044	127	158	-20	-40	BB3	1	4	273/12:25	025/07:07	88	24
93	22A206	168	187	-20	-32	BB3	1	4	285/00:23	036/11:49	169	59
94	22B044	168	187	-20	-32	BB3	1	4	285/00:23	036/11:49	169	59
95	22A119	168	187	-20	-32	BB3	1	4	164/16:48	016/17:26	273	25
96	22B000	43	73	+5	-10	BB4	1	4	279/15:56	031/06:40	84	19
97	22A118	43	73	+5	-10	BB4	1	4	264/16:40	016/17:18	272	26
98	21B054	122	161	0	-20	BB3	1	4	286/20:53	038/07:00	88	21
99	22A163	128	157	-20	-34	BB2	1	4	269/20:35	021/17:55	278	19
100	21A024	270	298	-10	-30	BB3	1	4	251/08:37	003/17:49	278	22
101	21A024	284	289	-15	-20	BB3	1	4	251/08:37	003/17:49	278	22
102	21A024	278	285	-15	-20	BB3	1	4	251/08:37	003/17:49	278	22
103	22A118	10	38	+10	-10	BB4	1	4	264/16 40	016/17:18	272	26
104	22B002	28	57	0	-20	BB3	1	4	279/16 37	031/07:21	91	25
105	22A236	60	90	0	-20	BB4	1	4	277/06:45	029/06:45	84	20
106	22B002	61	89	0	-20	BB3	1	4	280/22:30	032/12:35	189	60
107	21B008	45	75	+8	-10	BB4	1	4	280/16:26	032/06:30	82	17
108	21A237	45	75	+8	-10	BB4	1	4	277/20:02	029/12:06	175	60
109	21A120	29	115	+30	-30	SURV	1	4	264/16:52	016/17:30	274	24
110	22B003	272	335	+10	-5	BB4	1	4	279/21:18	031/12:02	174	60
111	22A135	272	335	+10	-5	BB4	1	4	266/18:17	018/17:35	275	22
112	21A108	30	65	0	-20	BB4	1	4	263/13:32	015/14:50	240	50
113	21A237	30	51	0	-10	BB4	1	4	277/20:02	029/12:06	175	60
114	22A207	275	306	0	-20	BB3	1	4	273/1233	025/07:15	89	25
115	22A133	287	312	0	-20	BB3	1	4	266/18:06	018/17:25	272	25
116	22A178	80	111	-20	-40	BB2	1	4	271/11:09	023/07:10	88	26
117	22A005	80	111	-20	-40	BB2	1	4	248/06:19	000/17:30	275	26
118	21B022	238	274	-20	-40	BB3	1	4	281/23:07	033/12:32	188	60
119	21A044	238	274	-20	-40	BB2	1	4	254/10:16	006/17:30	274	25
120	22B044	153	170	-20	-32	BB3	1	4	285/00:23	036/11:49	169	59
121	22A103	158	174	-20	-32	BB2	1	4	262/15:33	014/17:30	274	24

Figure Number	Camera Event Label	Azimuth start	Azimuth stop	Elevation top	Elevation bottom	Diode	Offset	Gain	Greenwich Mean Time	Local Lander Time	Sun azimuth	Sun elevation
122	21B021	185	271	-30	-50	BB1	1	4	281/22:32	033/11:57	172	60
123	21A132	253	282	-40	-60	BB1	1	4	266/18:01	018/17:20	272	25
124	21B033	214	249	-40	-60	BB1	1	4	282/23:44	034/12:30	187	60
125	21A031	214	249	-40	-60	BB1	1	4	252/08:27	004/17:00	269	30
126	22B058	63	92	-40	-60	BB1	1	4	287/02:15	038/12:22	183	59
127	21B040	180	222	-20	-40	BB2	1	4	284/19:35	036/07:01	88	22
128	21A052	185	222	-20	-40	BB2	1	4	255/09:26	007/16:00	257	40
129	21B040	222	262	-20	-40	BB2	1	4	284/19:35	036/07:01	88	22
130	21B059	150	190	0	-20	BB3	1	4	287/07:23	038/17:30	272	21
131	21B054	155	193	0	-20	BB3	1	4	286/20:53	038/07:00	88	21
132	21B034	156	185	0	-20	BB3	1	4	283/23:46	035/11:52	170	60
133	21A225	212	242	-10	-30	BB3	1	4	276/19:25	028/12:07	176	60
134	22A095	93	126	-40	-60	BB1	1	3	262/04:08	014/06:06	79	15
135	22B215	81	114	-40	-60	BB1	1	3	300/16:13	051/17:45	173	17
136	22A252	119	151	0	-20	BB4	1	4	278/20:43	030/12:06	175	60
137	22A252	151	215	0	-20	BB4	1	4	278/20:43	030/12:06	175	60
138	22A220	128	165	-4	-20	BB4	1	4	275/13:38	027/07:00	87	22
139	22A252	128	165	-4	-20	BB4	1	4	278/20:43	030/12:06	175	60
140	22A148	128	165	-4	-20	BB4	1	4	268/19:30	020/17:30	273	24
141	12A007	170	170	+30	-30	all	1	2	203/09:07	001/12:48	-	-
142	12A042	0	5	+30	-30	BB1	1	4	208/12:02	006/12:25	-	-
143	12A044	0	5	+30	-30	BB1	1	4	208/12:12	006/12:35	-	-
144	12A043	305	310	-10	-30	BB1	1	4	208/12:07	006/12:30	310	87
145	11A077	180	240	0	-20	BB4	1	4	213/18:22	011/15:27	282	48
146	11A077	180	240	0	-20	BB4	1	4	213/18:22	011/15:27	282	48
147	11A077	180	240	0	-20	BB4	1	4	213/18:22	011/15:27	282	48
148	12A004	132	150	0	-60	SURV	1	4	203/06:56	001/10:36	79	67
149	12A012	132	150	0	-60	SURV	1	4	204/05:00	002/08:00	74	33
150	12A026	132	150	0	-60	SURV	1	4	206/07:19	004/09:00	77	46
151	12A092	132	150	0	-60	SURV	1	4	215/13:21	013/09:06	78	48
152	11A037	214	217	-25	-40	BB2	1	4	207/14:41	005/15:43	1	28
153	21A078	245	245	-36	-50	BB1	1	4	259/10:34	011/14:30	235	53

Figure Number	Camera Event Label	Azimuth start	Azimuth stop	Elevation top	Elevation bottom	Diode	Offset	Gain	Greenwich Mean Time	Local Lander Time	Sun azimuth	Sun elevation
154	22B046	86	119	-20	-40	BB2	1	4	286/01:00	037/11:46	167	59
155	21B022	276	309	-20	-40	BB3	1	4	281/23:07	033/12:32	188	60
156	22D019	80	124	-20	-40	BB3	1	3	075/04:51	188/12:00	177	26
157	11A055	208	217	-25	-40	BB2	1	4	210/07:45	008/06:48	70	18
158	11A065	208	217	-25	-40	BB2	1	4	210/15:17	008/14:20	280	63
159	11B058	210	225	-25	-40	BB2	1	4	239/09:40	036/14:15	276	64
160	11B180	210	225	-25	-40	BB2	1	4	295/16:47	091/09:05	91	47
161	11C085	199	219	-20	-40	BB2	1	3	049/03:56	207/15:42	243	28
162	22A005	115	125	-24	-40	BB2	1	4	248/06:19	000/17:30	275	26
163	22A007	115	125	-24	-40	BB2	1	4	248/22:57	001/09:29	80	128
164	22A242	115	125	-24	-40	BB2	1	4	277/22:17	029/14:20	230	52
165	21A115	210	220	-20	-36	BB2	1	4	264/08:57	016/09:34	117	48
166	21B119	210	220	-20	-36	BB2	1	4	294/04:56	045/10:25	137	51
167	21B121	210	220	-20	-36	BB2	1	4	294/05:03	045/10:33	140	52
168	11B088	35	65	+10	-40	CLR	1	4	243/10:02	040/11:59	90	86
169	12B166	280	310	+10	-40	CLR	1	5	280/07:49	076/10:00	93	60
170	11B088	35	65	+10	-40	CLR	1	4	243/10:02	040/11:59	90	86
171	12B166	280	280	+10	-40	CLR	1	5	280/07:49	076/10:00	93	60
172	21C099	158	178	0	-20	BB3	1	4	015/11:35	130/09:00	129	26
173	12C087	247	249	+32	+27	SUN	1	1	049/04:59	207/16:44	251	15
174	11C105	193	195	+25	+20	SUN	1	1	051/22:35	210/08:22	117	28
175	11C177	237	239	+30	+25	SUN	1	1	082/18:56	240/08:55	126	28
176	11A088	182	187	+15	+10	SUN	1	2	215/10:03	013/05:48	66	5
177	11A089	185	190	+15	+10	SUN	1	2	215/10:27	013/06:12	68	10
178	11A090	185	193	+25	+20	SUN	1	2	215/11:15	013/07:00	71	20
179	12A176	55	60	+25	+19	BLU	1	2	231/11:59	028/21:51	316	-30
180	12F125	35	35	+20	-40	CLR	1	3	267/01:27	419/17:20	246	27
181	11A147	208	304	0	-60	CLR	1	4	229/00:33	026/11:44	82	83
182	11A147	208	304	0	-60	CLR	1	4	229/00:33	026/11:44	82	83
183	11A147	208	304	0	-60	CLR	1	4	229/00:33	026/11:44	82	83
184	11B088	35	65	+20	-40	CLR	1	4	243/10:02	040/11:59	90	86
185	11B042	175	205	0	-60	CLR	1	4	238/06:45	035/12:00	84	86
186	12A006	80	147	+10	-50	CLR	1	5	203/09:01	001/12:42	299	84
187	11B045	115	175	+30	-30	CLR	1	4	238/06:57	035/12:12	73	87
188	11B045	115	175	+30	-30	CLR	1	4	238/06:57	035/12:12	73	87
189	11A038	15	40	+10	-40	BLU	1	5	208/06:55	006/07:18	72	24
190	11A038	15	40	+10	-40	GRN	1	5	208/06:55	006/07:18	72	24

Figure Number	Camera Event Label	Azimuth start	Azimuth stop	Elevation top	Elevation bottom	Diode	Offset	Gain	Greenwich Mean Time	Local Lander Time	Sun azimuth	Sun elevation
191	11A038	15	40	+10	-40	RED	1	5	208/06:55	006/07:18	72	24
192	11A038	15	40	+10	-40	CLR	1	5	208/06:55	006/07:18	72	24
193	22A158	227	327	+30	-30	CLR	1	4	269/14:57	021/12:17	180	61
194	21A213	210	230	+10	-50	BLU	1	4	273/17:41	025/12:23	183	61
195	21E153	210	230	+10	-50	BLU	1	3	257/02:37	365/12:59	170	31
196	22E169	285	310	+20	-40	CLR	1	4	269/10:05	377/12:32	162	32
197	22A153	102	177	0	-60	CLR	1	4	269/12:10	021/09:30	117	47
198	22B133	105	132	-24	-46	BLU	1	4	295/07:01	046/11:51	170	57
	22B123	105	132	-24	-46	GRN	1	4	294/06:25	045/11:55	172	58
	22B115	105	132	-24	-46	RED	1	4	293/05:49	044/11:59	174	58
199	22A005	81	118	-20	-40	BB2	1	4	248/06:19	000/07:30	275	26
	22A007	81	118	-20	-40	BB2	1	4	248/22:57	001/09:29	80	128
200	12A239	130	190	+10	-50	CLR	1	0	233/10:16	030/18:48	293	4
201	22B023	110	170	+40	-20	CLR	1	5	282/15:44	034/04:29	61	2
202	12A240	140	250	+30	-30	CLR	1	5	233/10:41	030/19:13	295	-1
203	12A240	140	250	+30	-30	CLR	1	5	233/10:41	030/19:13	295	-1
204 to 226: Information not applicable												

The Authors

The success of the Viking mission represents the contributions of many thousands of persons whose signatures are contained in a microdot attached to each of the Viking Landers. Also, many people contributed directly to the Lander Camera Investigation. Some participated in the design and construction of cameras. Others planned the imaging sequences during the mission. Still others processed the imaging data returned to Earth. All should be identified as coauthors of this volume.

Some of these persons are identified below, according to their functional position. We say some because no listing, however long, is complete. In balance, we felt it more appropriate to list persons by name rather than to take refuge in the more common generalized acknowledgment to "all those who made this possible."

Viking Project Administrators

James S. Martin, Jr.
NASA Langley Research Center
Project Manager

Israel Taback
NASA Langley Research Center
Deputy Project Manager (Technical)

Gerald A. Soffen
NASA Langley Research Center
Project Scientist

A. Thomas Young
NASA Langley Research Center
Mission Director

G. Calvin Broome
NASA Langley Research Center
Lander Science Instruments Manager

Lander Imaging Science Team

Thomas A. Mutch
Brown University
Team Leader

Elliott C. Levinthal
Stanford University
Deputy Team Leader

Glenn R. Taylor
NASA Langley Research Center
Assistant Team Leader

Thomas A. Mutch
Kenneth L. Jones

William R. Patterson
Brown University
Team Engineer

Alan B. Binder
University of Kiel

Friedrich O. Huck
NASA Langley Research Center

Sidney Liebes, Jr.
Stanford University

Elliot C. Morris
U.S. Geological Survey

James B. Pollack
NASA Ames Research Center

Carl Sagan
Cornell University

Andrew T. Young
Texas A. & M. University

Lander Imaging Flight Team

All members of the Science Team plus the following:

Raymond E. Arvidson
Washington University
Also Science Team Leader, Extended Mission

Philip Avrin
Martin Marietta

C. Ernest Carlston
Martin Marietta

Robert D. Collie
NASA Langley Research Center

Paul Fox
Cornell University

Sven U. Grenander
Brown University

Kenneth L. Jones
Brown University

Carroll W. Rowland
NASA Langley Research Center

R. Stephen Saunders
Jet Propulsion Laboratory

Robert B. Tucker
Stanford University

Stephen D. Wall
NASA Langley Research Center

Michael R. Wolf
Jet Propulsion Laboratory

Camera Design, Construction, and Testing

Charles H. Ross
ITEK
Manager, camera construction program

Gerald Amante
EGG Incorp., Bedford Div.
Photosensor array fabrication

Antony Aponik
EGG Incorp., Bedford Div.
Photosensor array fabrication

Raymond P. Babcock
ITEK
Camera construction, Optical Alignment and Test
Equipment Engineer

Clarence J. Biggar
ITEK
Camera construction, electrical assembly

Michael D. Borenstein
Martin Marietta
Photosensor assembly, system level test

Troy G. Brooks
NASA Langley Research Center
Photosensor fabrication

John D. Buckley
NASA Langley Research Center
Photosensor fabrication

Ernest E. Burcher
NASA Langley Research Center
Facsimile camera design

Raymond J. Capobianco
ITEK
Camera construction, system quality assurance

Stephen L. Carman
Martin Marietta
Lander level camera testing

Vincent H. Corbett
Martin Marietta
ITEK Resident Team Manager

Parker J. Crossley
ITEK
Camera construction, program configuration

Billy B. Dancy
NASA Langley Research Center
Photosensor fabrication

William J. Davis
ITEK
Camera construction, Optical System Engineer

Donald K. Dean
Martin Marietta
ITEK Resident Team, quality assurance

Douglas P. Diederich
Martin Marietta
Camera testing

Roy A. Deveau
ITEK
Camera construction, Design Draftsman

Jules Dishler
Martin Marietta
Camera system requirements definition

John J. Donahue, Jr.
ITEK
Camera construction, systems engineering and Science Team liaison

Herbert W. Feinstein
ITEK
Camera construction, Staff Engineer for elevation axis – mechanical design

Joseph J. Fiorilla
ITEK
Camera construction, Program Chief Engineer

John C. Fleming
Martin Marietta
Photosensor assembly, filter design and testing

Kim B. Franzoni
ITEK
Camera construction, Contract Administrator

Robert N. Fuller, Jr.
ITEK
Camera construction, elevation and azimuth control system development

John W. Gerdes
ITEK
Camera construction, Tiger Team Program Manager

Norman S. Goralnick
ITEK
Camera construction, Development Engineer

Ray D. Harrell
Martin Marietta
Photosensor assembly, configuration management

Morriss L. Holliday
NASA Langley Research Center
Photosensor fabrication

Frederick H. Hudoff
Martin Marietta
Photosensor assembly, Program Manager

Kenneth L. Jacobs
NASA Langley Research Center
Consultant, servo design

Daniel J. Jobson
NASA Langley Research Center
Facsimile camera design

Michael Kardos
Martin Marietta
ITEK Resident Team, Product Integrity Engineer

Stephen J. Katzberg
NASA Langley Research Center
Facsimile camera design

W. Lane Kelly IV
NASA Langley Research Center
Facsimile camera design

Melvin W. Kuethe
Martin Marietta
Dust protection design and testing

Jules J. Lambiotte, Jr
NASA Langley Research Center
Facsimile camera design

Frank L. Lawrence
NASA Langley Research Center
Photosensor fabrication

Lawrence Leonard
EGG Incorp., Bedford Div.
Photosensor array fabrication

William F. Loughlin
ITEK
Camera construction, electronics assembly integration

Samuel C. Lukes
Martin Marietta
CO_2 dust removal, Program Manager

Thomas K. Lusby, Jr.
NASA Langley Research Center
Photosensor fabrication

Allan J. MacEachern
ITEK
Camera construction, Tiger Team Chief Planner

James W. MacFarlane
ITEK
Ground reconstruction equipment, Project Engineer

Clifford J. Maxwell
ITEK
Ground reconstruction equipment

Henry F. McCall
ITEK
Camera construction, Electronics Lead Engineer

John H. Montgomery
Martin Marietta
Camera electronic design analysis

Raymond Peluso
Martin Marietta
Photosensor array fabrication

Robert B. Penninger
ITEK
Ground reconstruction equipment, Program Engineer

Albert R. Philips
NASA Langley Research Center
Photosensor fabrication

John J. Rezendes
ITEK
Camera construction, program production control

James H. Rich
ITEK
Camera construction, mechanical manufacture,
Technical-Specialist

Kenneth H. Schlichtemeier
Martin Marietta
Camera electronics, worst case analysis

Rennald C. Schmidt
Martin Marietta
Photosensor assembly, quality control

Eric Schwarm
Martin Marietta
Camera servo design analysis

Donald A. Stang
Martin Marietta
Dust protection device, fabrication

Staulo Salvatore
ITEK
Camera construction, Chief Mechanical Engineer

Lyman A. Stilley, Jr.
NASA Langley Research Center
Photosensor fabrication

David Stout
Martin Marietta
Preliminary camera electronics design

Rush B. Studinski
Martin Marietta
ITEK Resident Team, Subcontract Administrator

Eugene R. Sullivan
ITEK
Camera construction, Program Administrator

Rowland Tully
NASA Langley Research Center
Photosensor fabrication

Donald L. Tushin
ITEK
Camera construction, component and system testing

Doris M. Tyler
ITEK
Camera construction, printed circuit card assembly

John G. Waring
Martin Marietta
Field test electronic design

Maywood L. Wilson
NASA Langley Research Center
Photosensor fabrication

Virgil F. Young
Martin Marietta
Photosensor assembly, circuit design

Image Formatting and Processing

Kermit S. Watkins
Jet Propulsion Laboratory
Manager, image processing system development

David L. Atwood
Jet Propulsion Laboratory
Calibration data processing

William J. Beckmann
Jet Propulsion Laboratory
Photolaboratory technician

William D. Benton
Jet Propulsion Laboratory
Color image reconstruction

Joseph W. Berry
Jet Propulsion Laboratory
Geometric rectification software, computer mosaics

D. Thomas Box
Jet Propulsion Laboratory
Supervisor, Mission and Test Photosystem

Roger Brandt
Jet Propulsion Laboratory
Computer hardware configuration and maintenance

William J. Chandler
Jet Propulsion Laboratory
Mission and Test Photosystem Engineer

Charles E. Frasier
Jet Propulsion Laboratory
First order image processing software

Joseph Fulton
Jet Propulsion Laboratory
Computer hardware configuration and maintenance

Joseph D. Gaunder
Jet Propulsion Laboratory
First order image processing

Michael A. Girard
Jet Propulsion Laboratory
Interactive image processing system software

William B. Green
Jet Propulsion Laboratory
Leader, image processing staff

Lawrence S. Green
Jet Propulsion Laboratory
First order image processing software

William Harlow
Stanford University
Stereo station construction

Paul Harari
Jet Propulsion Laboratory
First order image processing

John C. Hewitt
Jet Propulsion Laboratory
Supervisor, photolaboratory

Paul L. Jepson
Jet Propulsion Laboratory
Interactive image display system software

Ralph A. Johansen
Jet Propulsion Laboratory
Supervisor, Mission and test imaging system

Raymond Jordan
U.S. Geological Survey
Stereoplotter

Susan K. LaVoie
Jet Propulsion Laboratory
Team data record processing, computer mosaics

Edward Y. S. Lee
Jet Propulsion Laboratory
Interactive picture catalog system

Fred H. Lesh
Jet Propulsion Laboratory
System Programmer

Jean J. Lorre
Jet Propulsion Laboratory
Radiometric decalibration software

Donald J. Lynn
Jet Propulsion Laboratory
Deputy Leader, image processing staff

Michael D. Martin
Jet Propulsion Laboratory
Photoproduct storage and distribution system

Mary G. McGough
Jet Propulsion Laboratory
Photolaboratory Technician

Edward Morita
Jet Propulsion Laboratory
First order image processing

Duane K. Patterson
Jet Propulsion Laboratory
Photolaboratory Technician

Rodger N. Philips
Jet Propulsion Laboratory
Data logging, graphics display software

Robert W. Post
Jet Propulsion Laboratory
Photolaboratory Technician

Margaret A. Power
Jet Propulsion Laboratory
Stereopair processing and projection software

Gerald A. Praver
Jet Propulsion Laboratory
Manager, Mission and test photosystem

Lynn Quam
Stanford University
Image processing software

Arnold A. Schwartz
Jet Propulsion Laboratory
Lander stereophotogrammetry analysis and software

Joel B. Seidman
Jet Propulsion Laboratory
Supervisor, Image Processing Laboratory Operations Group

Ted Sesplaukis
Jet Propulsion Laboratory
Automated image cataloguing

Steven Silverman
Jet Propulsion Laboratory
Computer hardware configuration and maintenance

Deborah M. Spurlock
Jet Propulsion Laboratory
Experiment data record production

Norma J. Stetzel
Jet Propulsion Laboratory
Mission and Test Photosystem Quality Control Engineer

V. W. Tuk
Jet Propulsion Laboratory
Computer hardware configuration and maintenance

Jurrie J. Van der Woude
Jet Propulsion Laboratory
Photolaboratory Group Leader

Ronald J. Wichelman
Jet Propulsion Laboratory
Photographic engineer

Jackson Wilson
Jet Propulsion Laboratory
Automated image cataloguing

Sherman S. C. Wu
U.S. Geological Survey
Photogrammetry

Mission Planning

Kayland Z. Bradford
Martin Marietta
Image sequence simulation

Elizabeth F. Buchan
NASA Langley Research Center
Image sequencing

Carolyn G. Cooley
Martin Marietta
Image sequence planning

William C. Eggemeyer
Washington University
Image sequencing

R. Terry Gamber
Martin Marietta
Image sequence planning

Lloyd E. Gilbert
Martin Marietta
Imaging software development

James Gliozzi
Martin Marietta
Sample site imaging coordination

Jaryl K. Kerekes
Martin Marietta
Image sequencing

B. Gentry Lee
Martin Marietta
Director, Science Analysis and Mission Planning

Donald W. Marquet
Martin Marietta
Mission event sequence planning

Deborah G. Pidek
Martin Marietta
Image sequencing

James D. Porter
Martin Marietta
Mission event sequence planning

Albert R. Schallenmuller
Martin Marietta
Mission software development and integration

Janet A. Shields
Martin Marietta
Image sequencing

Henry C. Von Struve III
Martin Marietta
Lander science planning

Edward J. Taylor
General Electric Corporation
Imaging planning and software development

Hugh N. Zeiner
Martin Marietta
Image sequence planning

Adjunct Science Analysts

Edward W. Dunham
Cornell University

Edward A. Guinness
Washington University

Bruce W. Hapke
University of Pittsburgh

James W. Head 111
Brown University

Ralph A. Kahn
Harvard University

Baerbel K. Lucchitta
U.S. Geological Survey

Dag Nummedal
University of South Carolina

David C. Pieri
Cornell University

James R. Underwood, Jr.
Kansas State University

Support Personnel

Sarah H. Bosworth
Brown University
Secretary

Richard E. D'Alli
Brown University
Viking Intern Program Administrator

Leslie A. Lloyd
Jet Propulsion Laboratory
Photolibrarian

F. Wayne Loer
NASA Langley Research Center
Secretary

Madeline M. Mutch
Providence, R.l.
Sustainer

Charlotte M. Scherig
ITEK
Secretary

Viking Interns
Undergraduates selected by national competition to participate in mission operations

Richard O. Ackermann
California Institute of Technology

Steven Albers
Muhlenberg College

Kenneth G. Carpenter
Wesleyan University

Andrew L. Chaikin
Brown University

Eric H. Christiansen
Brigham Young University

R. Glenn Cooper
Rice University

Mark A. Croom
Virginia Polytechnic Institute

William E. Dieterle
Arizona State University

Mark B. DuBois
Harvard University

Chris Eberspacher
University of Texas

F. Donald Eckelmann, Jr.
Colgate University

Mark I. Heiligman
Amherst College

Edward A. Hildum
Massachusetts Institute of Technology

Albert R. Hochevar
U.S. Naval Academy

Julio Magalhaes
Stanford University

William B. McKinnon
Massachusetts Institute of Technology

Kurt D. Molenaar
Macalester College

Barbara A. Murphy
Lehigh University

John J. Polcari
U.S. Naval Academy

David W. Thompson
Massachusetts Institute of Technology

Judy C. Thompson
Massachusetts Institute of Technology

Michael C. Wright
University of Tennessee

www.ingramcontent.com/pod-product-compliance
Lightning Source LLC
Chambersburg PA
CBHW081505200326
41518CB00015B/2389

* 9 7 8 1 7 8 2 6 6 4 8 8 8 *